Volume Two
Student Solutions
Manual for Reese's
University Physics

Ronald Lane Reese
Washington & Lee University

Robin B. S. Brooks
Bates College

Mark D. Semon
Bates College

Brooks/Cole
Thomson Learning.

Pacific Grove • Albany • Belmont • Boston • Cincinnati • Johannesburg • London • Madrid
Melbourne • Mexico City • New York • Scottsdale • Singapore • Tokyo • Toronto

Sponsoring Editor: *Melissa D. Henderson*
Editorial Assistant: *Dena Dowsett-Jones*
Marketing Manager: *Steve Catalano*
Marketing Assistant: *Christina De Veto*

Production Coordinator: *Dorothy Bell*
Cover Design: *Roy R. Neuhaus*
Cover Photo: *Ronald Lane Reese*
Printing and Binding: *Globus Printing*

Printed in the United States of America

10   9   8   7   6   5   4   3   2   1

ISBN 0-534-37021-7

Contents

Preface

This solutions manual is for you. One of the best ways to learn is through apprenticeship; however, since we can't work side by side with you, we offer this book in our place. We hope reading it will be like looking over our shoulders as we solve the problems, and that you will benefit from seeing how we, as more experienced problem solvers, proceed.

Volume One of the *Student Solutions Manual* contains solutions to every other odd problem $(1, 5, 9, 13, \dots)$ in Chapters 1 through 15 of *University Physics*; Volume Two contains solutions to every other odd problem in Chapters 16 through 27. For your convenience, problems in *University Physics* are numbered in red if their solutions are in either Volume One or Volume Two of the *Student Solutions Manual*.

More often than not, there are several different ways of solving a problem (see page 12 of *University Physics* for a discussion of this), so it's all right if your way differs from ours – as long as your method is correct! The most beneficial way of using this book is to attempt the problem yourself first, and only then to look at how we solve it. It is one thing to solve a problem yourself and a totally different thing to read someone else's solution. If you read our solution first, hopefully you will think: this makes sense, that makes sense, and so on; but then, chances are you will not be able to solve the problem when you encounter it on a test or in some other situation in which you can't refer to the solutions manual. We don't think you want to find yourself in that position! On the other hand, if you try the problem *before* reading our solution and get stuck, you create in yourself a "need to know" which makes you more likely to truly grasp the point(s) you didn't fully understand.

Physics is not a spectator sport. Just as listening to a piano concerto is not the same as playing it yourself, watching (or reading) how someone else solves problems won't improve *your* ability to handle them. Piano playing and problem solving both take consistent practice to master.

Although the problems in *University Physics* span a wide variety of subjects, there are some common elements in how we suggest you approach them. (See page 12 of *University Physics* for an elaboration on this.) First, whenever possible, make a sketch of the situation involved. Second, identify which quantities you know and which you wish to know. Third, review the laws and equations involving these quantities to see how what you wish to know is related to what you already know. Also, pay attention to the Problem – solving Tactics discussed in each chapter of the text; they are there to help you solve different types of problems and avoid common pitfalls along the way. All this may sound simple, but most of us need a lot of practice to really learn it, and to develop the conviction that the answers we seek really are contained in the laws we have learned – no matter how unlikely this may seem at first glance!

In some cases your final numerical solution may differ slightly from ours. This is most likely due to the number of 'significant digits' you carry through your calculations. We have expressed our answers to the precision permitted by the numbers upon which they are based, so if your answer differs slightly from ours, be sure to check whether you are using the proper number of significant figures. (See pages 18-21 of *University Physics* for a more complete discussion of this.) In some cases, however, an answer will depend upon where the correct number of significant digits is imposed. For example, in problems with several parts, we quote the answer to each part with the proper number of significant digits and calculate what follows using this number. Because of this, answers in the parts that follow might differ from what you would have found if you hadn't stopped along the way to quote an intermediate result. If you prefer to complete the whole problem before imposing the correct number of significant figures, don't worry if your answer differs slightly from ours.

Also, different people have different 'rounding' conventions. For example, in a situation where only two significant digits are allowed, some people will round a number like 7.65 up to 7.7 while others will round it down to 7.6. We have chosen the former convention, and in fact, we round all numbers *up* whose first 'insignificant digit' is a five, regardless of whether the digit preceding it is even or odd.

We have made an effort to keep our notation and terminology consistent with the text. There may, however be a few small differences, in particular we use a bold face zero **0** to denote the zero vector, while the text uses the same notation, 0, for both the scalar and the zero vector.

One of the most common statements a teacher hears is "I understand the material but I just can't do the problems." Although problem-solving is a skill in its own right, we believe that if you cannot work the problems you do not fully understand the material. You might *know* the material, but you have not really *understood* it. In our view, *understanding* is the result of *knowledge and experience*, and *experience* only comes from *using* the material in hands-on activities, like solving problems and doing laboratory experiments.

Finally, we want to encourage you heartily! Best wishes for success on your journey into physics, engineering, mathematics, and other sciences. It is not an easy road, but it is one we have found to be intellectually rewarding and enriching. Some people (erroneously) believe that learning science is memorizing a large collection of facts and formulas. We hope you will see how miraculous it is that most of these facts and formulas can be understood within the framework of remarkably few laws of nature. We also hope that in solving these problems, you experience the bewilderment of seeing something you have not seen before, the thrill of understanding it, and the wonder of seeing how it all fits together into a bigger picture.

We had a good time doing these problems, and we hope you do too!

Ronald Lane Reese
Dept. of Physics
Washington and Lee University
Lexington, VA 24450
reeser@wlu.edu

Robin B.S. Brooks
Dept. of Mathematics
Bates College
Lewiston, ME 04240
rbrooks@bates.edu

Mark D. Semon
Dept. of Physics
Bates College
Lewiston, ME 04240
msemon@bates.edu

25 January 1999

Acknowledgments

We thank our editors at Brooks/Cole for all their help. In particular, we thank Senior Assistant Editor Melissa Henderson, Physics Editor Beth Wilbur, and Senior Developmental Editor Keith Dodson for their encouragement, support, and dedication to excellence.

We also thank our "Checkers," the many anonymous physicists who rechecked our calculations to ensure accuracy, and who discovered typographical errors in what we *thought* was perfect copy.

This solutions manual was typeset entirely on a personal computer using LaTeX, an extension of TeX. The TeX program and language were created by Donald Knuth. We thank Professor Knuth and the many people following him for their marvelous gift to the world.

Chapter 16

Electrical Charges, Electrical Forces, and the Electric Field

16.1 A mole has Avogadro's number of particles. Hence the magnitude of charge on a mole of electrons is

$$N_A e = (6.0221 \times 10^{23} \text{ mol}^{-1})(1.6022 \times 10^{-19} \text{ C}) = 9.6486 \times 10^4 \text{ C/mol}.$$

16.5 Each hydrogen atom has one proton with charge e, so the total positive charge in a mole of hydrogen atoms is

$$N_A e = (6.0221 \times 10^{23} \text{ mol}^{-1})(1.6022 \times 10^{-19} \text{ C}) = 9.6486 \times 10^4 \text{ C/mol}.$$

16.9

a) Find the magnitude of the electrical force of the gold nucleus on the α-particle from Coulomb's law:

$$F_{\text{elec}} = \frac{1}{4\pi\epsilon_0} \frac{|q|\,|Q|}{r^2} = (9.00 \times 10^9 \text{ N·m}^2/\text{C}^2)\frac{2(1.602 \times 10^{-19} \text{ C})\,79(1.602 \times 10^{-19} \text{ C})}{(1.00 \times 10^{-14} \text{ m})^2} = 365 \text{ N}.$$

b) The electrical forces of the gold nucleus and the α-particle on each other form a Newton's third law force pair. Hence the magnitude of the electrical force of the α-particle on the gold nucleus is also 365 N.

16.13 From Newton' second law, the total force on the Earth must equal its mass times its acceleration. Use the magnitude of the vectors. The force in this case is electrical, and the acceleration is the centripetal acceleration.

$$F = ma \implies \frac{1}{4\pi\epsilon_0} \frac{|q||q|}{r^2} = m\frac{v^2}{r},$$

so, after solving for q,

(1)
$$|q| = v\sqrt{\frac{mr}{\left(\dfrac{1}{4\pi\epsilon_0}\right)}}.$$

The speed of the Earth in its orbit is the circumference of the orbit (a distance) divided by its period T (time to traverse the distance), $v = \dfrac{2\pi r}{T}$. Substitute this expression for v into (1) to get

(2)
$$|q| = \frac{2\pi r}{T}\sqrt{\frac{mr}{\left(\dfrac{1}{4\pi\epsilon_0}\right)}}.$$

The period T is one year. Convert this to seconds.

$$T = 1\,\text{y} = (1\,\text{y})\left(\frac{365.25\,\text{d}}{\text{y}}\right)\left(\frac{8.6400 \times 10^4\,\text{s}}{\text{d}}\right) = 3.1558 \times 10^7\,\text{s}\,.$$

Substitute numerical values into (2).

$$|q| = \frac{2\pi(1.49 \times 10^{11}\,\text{m})}{3.1558 \times 10^7\,\text{s}}\sqrt{\frac{(5.98 \times 10^{24}\,\text{kg})(1.49 \times 10^{11}\,\text{m})}{9.00 \times 10^9\,\text{N·m}^2/\text{C}^2}} = 2.95 \times 10^{17}\,\text{C}\,.$$

Notice that in working this and similar problems, we find it more convenient to work with the numerical value of $\frac{1}{4\pi\epsilon_0} = 9.00 \times 10^9\,\text{N·m}^2/\text{C}^2$ than with the numerical value of $4\pi\epsilon_0$.

16.17

a)   The forces on each mass are:

   1. its weight $\vec{\mathbf{w}}$, of magnitude mg, directed down;

   2. the force $\vec{\mathbf{T}}$ of the cord, pointing away from the mass and along the cord; and

   3. the electrical force $\vec{\mathbf{F}}_{\text{elec}}$, pointing horizontally away from the other mass. (Since the charges are like charges, this force is repulsive.)

Here is the second law force diagram and an appropriate coordinate system.

b)   Consider just the mass on the left. In the above coordinate system, the forces on it are:

$$\vec{\mathbf{w}} = -mg\hat{\mathbf{j}}, \qquad \vec{\mathbf{T}} = T\sin\theta\,\hat{\mathbf{i}} + T\cos\theta\,\hat{\mathbf{j}}, \qquad \text{and} \qquad \vec{\mathbf{F}}_{\text{elec}} = -F_{\text{elec}}\hat{\mathbf{i}}\,.$$

The separation of the charges is $r = 2\ell\sin\theta$, so

$$F_{\text{elec}} = \frac{1}{4\pi\epsilon_0}\frac{|q||q|}{r^2} = \frac{1}{4\pi\epsilon_0}\frac{q^2}{4\ell^2\sin^2\theta}\,.$$

The mass is in static equilibrium, so the total force on it is zero. Thus, in the x direction we have

(1) $$T\sin\theta - \frac{1}{4\pi\epsilon_0}\frac{q^2}{4\ell^2\sin^2\theta} = 0\,\text{N}\,.$$

In the y direction we have

$$-mg + T\cos\theta = 0\,\text{N} \implies T = \frac{mg}{\cos\theta}\,.$$

Now substitute this expression for T into (1) and solve for q.

$$\frac{mg}{\cos\theta}\sin\theta - \frac{1}{4\pi\epsilon_0}\frac{q^2}{4\ell^2\sin^2\theta} = 0\,\text{N} \implies mg\tan\theta - \frac{1}{4\pi\epsilon_0}\frac{q^2}{4\ell^2\sin^2\theta} = 0\,\text{N}$$

$$\implies |q| = 2\ell\sin\theta\sqrt{4\pi\epsilon_0 mg\tan\theta}\,.$$

c)   Substitute the numerical values and change the mass units from grams to kilograms:

$$|q| = 2\ell \sin\theta \sqrt{4\pi\epsilon_0 mg \tan\theta} = 2\ell\sin\theta\sqrt{\dfrac{mg\tan\theta}{\left(\dfrac{1}{4\pi\epsilon_0}\right)}}$$

$$= 2(0.500\text{ m})\sin 15.0°\sqrt{\dfrac{(0.010\text{ kg})(9.81\text{ m/s}^2)\tan 15.0°}{9.00\times 10^9\text{ N·m}^2/\text{C}^2}} = 4.4\times 10^{-7}\text{ C}.$$

Notice that in working this and similar problems, we find it more convenient to work with the numerical value of $\dfrac{1}{4\pi\epsilon_0} = 9.00\times 10^9$ N·m^2/C^2 than with the numerical value of $4\pi\epsilon_0$.

d)   The charge quantum number n is found from

$$q = ne \implies n = \dfrac{q}{e} = \dfrac{\pm 4.4\times 10^{-7}\text{ C}}{1.602\times 10^{-19}\text{ C}} = \pm 2.7\times 10^{12}.$$

16.21   The position vector $\vec{r}_{\text{CM}}$ of the center of mass is

$$\vec{r}_{\text{CM}} \overset{\text{def}}{\equiv} \dfrac{\sum_i m_i\vec{r}_i}{\sum_i m_i},$$

where $\vec{r}_i$ is the position vector of a particle with mass m_i.

By analogy, the position vector of the center of charge is

$$\vec{r}_{\text{center of charge}} \overset{\text{def}}{\equiv} \dfrac{\sum_i q_i\vec{r}_i}{\sum_i q_i}$$

where $\vec{r}_i$ is the position vector of the particle with charge q_i.

A major difference is that, unlike the m_i, the q_i can be negative as well as positive, so it is possible to have a collection of particles each with nonzero charge, but with total charge $\sum_i q_i$ equal to zero. In this case the position vector of the center of charge is undefined and meaningless, for this condition produces a zero in the denominator of the defining equation.

16.25   The force on a charge q in an electric field $\vec{E}$ is

$$\vec{F}_{\text{elec}} = q\vec{E}.$$

Since the direction of the force on the professor is opposite to the direction of the field, the professor's charge q must be negative. The magnitude of the force is

$$F_{\text{elec}} = |q|E \implies 6.0\text{ N} = |q|500\text{ N/C} \implies |q| = 1.2\times 10^{-2}\text{ C}.$$

So, since q is negative, $q = -1.2\times 10^{-2}$ C.

16.29

a)   Since the sphere is being deflected in a direction opposite to the direction of the electric field, the charge on it must be negative.

b)   The forces acting on the small charged sphere are:

1. its weight $\vec{w}$, of magnitude mg, directed down;

2. the tension $\vec{T}$ of the string, directed along the string and away from the sphere; and

3. the force $\vec{F}_{\text{elec}}$ of magnitude $|q|E$, directed horizontally to the right.

Here's the second law force diagram, together with an appropriate choice of coordinate system.

c)   Using the coordinate system in the above sketch, the forces on the sphere are:

$$\vec{\mathbf{w}} = -mg\hat{\mathbf{j}}, \qquad \vec{\mathbf{T}} = -T\sin\theta\hat{\mathbf{i}} + T\cos\theta\hat{\mathbf{j}}, \qquad \text{and} \qquad \vec{\mathbf{F}}_{\text{elec}} = q(-E\hat{\mathbf{i}}),$$

where $m = 0.0200$ kg is the mass of the sphere, and q is the unknown (but negative) charge on the sphere.

Since the charge is at rest, Newton's second law implies that the total force on the mass is zero. Thus, in the x direction we have

(1) $$-T\sin\theta - qE = 0 \text{ N} \implies q = \frac{-T\sin\theta}{E}.$$

In the y direction we have

$$-mg + T\cos\theta = 0 \text{ N} \implies T = \frac{mg}{\cos\theta}.$$

Now sustitute this expression for T into (1) and compute.

$$q = \frac{-\dfrac{mg}{\cos\theta}\sin\theta}{E} = \frac{-mg\tan\theta}{E} = \frac{-(0.0200 \text{ kg})(9.81 \text{ m/s}^2)\tan 10^\circ}{100 \text{ N/C}} = -3.5 \times 10^{-4} \text{ C}.$$

16.33   Electric field lines are directed away from positive charges and toward negative charges. Hence, the charge on the far left is positive while the other two are negative. Notice that the number of field lines from the two negative charges are the same, so these charges have equal magnitude. The number of field lines leaving the positive charge is twice those approaching either negative charge. Hence the positive charge has a magnitude twice that of either negative charge. The charges are $2q$, $-q$, and $-q$, where q is positive.

16.37   The information implies that the three atoms of the molecule are collinear. The dipole moments of the two carbon-oxygen bonds therefore point in opposite directions and vector sum to zero, giving the molecule as a whole a dipole moment equal to **0** C·m.

16.41

a)   From Figure 16.63 on page 729 of the text, the approximate size of the water molecule is 0.1 nm. The charge Q is 10.0 nm from the dipole, a distance approximately 100 times the size of the dipole. Hence, the two charges of the dipole experience approximately the same field $\vec{\mathbf{E}}$, where $\vec{\mathbf{E}}$ is the electric field produced by Q. The total force on the dipole is therefore

$$\vec{\mathbf{F}} = (+q)\vec{\mathbf{E}} + (-q)\vec{\mathbf{E}} = \mathbf{0} \text{ N}.$$

b)   The electric field $\vec{\mathbf{E}}$ produced by Q at the position of the dipole is pointed toward Q, because Q is negative. According to the problem, the dipole $\vec{\mathbf{p}}$ is also pointed toward Q. Therefore at the position of the dipole, $\vec{\mathbf{p}}$ and $\vec{\mathbf{E}}$ are *parallel*, so the torque is

$$\vec{\tau} = \vec{\mathbf{p}} \times \vec{\mathbf{E}} = \mathbf{0} \text{ N·m}.$$

The torque is zero.

16.45   Use the coordinate system sketched below.

The torque on the dipole is $\vec{\tau} = \vec{p} \times \vec{E}$. In the above coordinate system, and using the right-hand rule, $\vec{p} \times \vec{E}$ is a vector with direction $-\hat{k}$ and magnitude $pE\sin\theta$. Hence

$$\vec{\tau} = -pE\sin\theta\,\hat{k}.$$

Substitute this expression for $\vec{\tau}$ in the equation $\vec{\tau} = I\vec{\alpha}$, to get

(1) $-pE\sin\theta\,\hat{k} = I\vec{\alpha}$

where $\vec{\alpha}$ is the angular acceleration of the dipole. Notice from this expression that although the magnitude of the angular acceleration can change with time, it is always in the direction $\pm\hat{k}$. Thus, if at any instant the angular velocity $\vec{\omega}$ is a scalar multiple of $\hat{k}$, then it will always be a scalar multiple of $\hat{k}$. This would happen, for example, if the dipole had been originally "released from rest," for then, at the time of release, its angular velocity would have been $(0 \text{ rad/s})\hat{k}$. We will therefore assume that the angular velocity is a scalar multiple of $\hat{k}$. Then the angular velocity is just $\frac{d\theta}{dt}\hat{k}$, so the angular acceleration is

$$\vec{\alpha} = \frac{d}{dt}\left(\frac{d\theta}{dt}\hat{k}\right) = \frac{d^2\theta}{dt^2}\hat{k}.$$

Now use this expression for $\vec{\alpha}$, to rewrite (1) as

$$-pE\sin\theta\,\hat{k} = I\frac{d^2\theta}{dt^2}\hat{k} \implies \frac{d^2\theta}{dt^2} + \left(\frac{pE}{I}\right)\sin\theta = 0 \text{ rad/s}^2.$$

For small angles θ measured in radians, $\sin\theta \approx \theta$. Using this approximation as an equality in the last equation, we have

$$\frac{d^2\theta}{dt^2} + \left(\frac{pE}{I}\right)\theta = 0 \text{ rad/s}^2.$$

This is the differential equation for simple harmonic oscillations in θ. The coefficient of θ in the equation is the square of the angular frequency ω_{osc} of the oscillation, so

$$\omega_{\text{osc}} = \sqrt{\frac{pE}{I}}.$$

Therefore the frequency of the oscillation is

$$\nu = \frac{\omega_{\text{osc}}}{2\pi} = \frac{1}{2\pi}\sqrt{\frac{pE}{I}}.$$

16.49   Choose a coordinate system with $\hat{j}$ pointing up. The only forces on the cork are its weight, $-mg\hat{j}$, and the electric force $\vec{F}_{\text{elec}} = q\vec{E}$, where $m = 2.00 \times 10^{-3}$ kg is the mass of the cork and $q = 3.00 \times 10^{-6}$ C is the charge on the cork. The cork is not accelerating, so the total force on it is zero. Also $q > 0$ C. Hence

$$-mg\hat{j} + q\vec{E} = \mathbf{0} \text{ N} \implies mg = |q|E = qE.$$

The magnitude E of the electric field of a uniformly charged (infinite) sheet is

$$E = \frac{\sigma}{2\epsilon_0}.$$

Thus, substituting this expression for E and solving for σ,

$$mg = qE = q\frac{\sigma}{2\epsilon_0} \implies \sigma = 2\epsilon_0\left(\frac{mg}{q}\right) = \left(\frac{4\pi\epsilon_0}{2\pi}\right)\left(\frac{mg}{q}\right)$$

$$\implies \sigma = \frac{1}{(2\pi)(9.00 \times 10^9 \text{ N·m}^2/\text{C}^2)}\left(\frac{(2.00 \times 10^{-3} \text{ kg})(9.81 \text{ m/s}^2)}{3.00 \times 10^{-6} \text{ C}}\right) = 1.16 \times 10^{-7} \text{ C/m}^2.$$

16.53   From problem 16.52, the magnitude of the field at the center of a half-ring is

$$E = \frac{1}{4\pi\epsilon_0}\frac{2Q}{\pi R^2},$$

where R is the radius of the half-ring, and Q is its total charge. The length of the half-ring is given as $\ell = 0.500\ \text{m}$. Since $\ell = \pi R$, $R = \dfrac{\ell}{\pi}$, so in terms of of ℓ,

$$E = \frac{1}{4\pi\epsilon_0}\frac{2Q}{\pi\left(\dfrac{\ell}{\pi}\right)^2} = \frac{1}{4\pi\epsilon_0}\frac{2Q\pi}{\ell^2} = (9.00\times10^9\ \text{N·m}^2/\text{C}^2)\frac{2|-8.00\times10^{-9}\ \text{C}|\pi}{(0.500\ \text{m})^2} = 1.81\times10^3\ \text{N/C}.$$

Here, along with a convenient coordinate system, is how the field looks at the center of the half-ring.

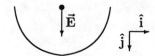

Since the half-ring is symmetric about the vertical through its center, the field along $\hat{\mathbf{i}}$ produced by the charge to the left of the vertical line cancels the field produced by the charges to the right of the line. Because the total charge on the ring is negative, the field points towards the ring, rather than away from it. In terms of the coordinate system

$$\vec{\mathbf{E}} = (1.81\times10^3\ \text{N/C})\hat{\mathbf{j}}$$

at the center of the ring.

16.57

a)   In order for the cord to be taut, the tiny mass and the infinite sheet should repel each other. This will occur if and only if they have like charges — both of them positive or both of them negative.

b)   When q is moved slightly away from its equilibrium position and then released with the cord still taut, the forces on q are:

1. the force $\vec{\mathbf{T}}$ of the cord, directed along the cord and away from q; and

2. the electrical force $\vec{\mathbf{F}}_{\text{elec}} = q\vec{\mathbf{E}}$, where $\vec{\mathbf{E}}$ is the electric field produced by the infinite sheet of charge.

Here's the second law force diagram together with an appropriate coordinate system.

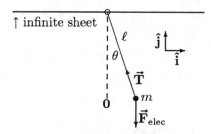

The origin **0** is at the equilibrium position for m. When m is at the origin the two forces $\vec{\mathbf{T}}$ and $\vec{\mathbf{F}}_{\text{elec}}$ vector sum to zero, so the mass is not accelerating.

Notice that this situation is precisely analagous to the pendulum situation. The above diagram is essentially the same as Figure 7.23 on page 298 of Volume 1 of the text. The only difference is that here the electric force $\vec{\mathbf{F}}_{\text{elec}} = -qE\hat{\mathbf{j}}$ has replaced the gravitational force $\vec{\mathbf{w}} = -mg\hat{\mathbf{j}}$.

In this coordinate system, the forces on m are

$$\vec{\mathbf{T}} = -T\sin\theta\,\hat{\mathbf{i}} + T\cos\theta\,\hat{\mathbf{j}}, \quad \text{and} \quad \vec{\mathbf{F}}_{elec} = -qE\,\hat{\mathbf{j}}.$$

Just as in the analysis of the pendulum, we suppose the angle θ small enough so that we can neglect any acceleration in the y direction. Then in the y direction, the total force is zero, so

$$T\cos\theta - qE = 0 \text{ N} \implies T\cos\theta = qE.$$

Since θ is small, $\cos\theta \approx 1$, so we replace $\cos\theta$ by 1 and (bravely) rewrite this last equation as

(1) $$T = qE.$$

Now apply Newton's second law to the x direction,

$$-T\sin\theta = ma_x = m\frac{dx}{dt} \implies \frac{dx}{dt} + \left(\frac{T}{m}\right)\sin\theta = 0 \text{ m/s}^2.$$

From the geometry, $\dfrac{x}{\ell} = \sin\theta$, so the last equation becomes

$$\frac{dx}{dt} + \left(\frac{T}{m}\right)\left(\frac{x}{\ell}\right) = 0 \text{ m/s}^2.$$

Finally, use equation (1) to substitute qE for T, and then simplify.

$$\frac{dx}{dt} + \left(\frac{qE}{m\ell}\right)x = 0 \text{ m/s}^2.$$

This is the equation for simple harmonic oscillation. The coefficient of x is the square of the angular frequency, ω, of the oscillation, so

$$\omega = \sqrt{\frac{qE}{m\ell}}.$$

The frequency is

$$\nu = \frac{\omega}{2\pi} = \frac{1}{2\pi}\sqrt{\frac{qE}{m\ell}}.$$

The magnitude of the electric field of an infinite sheet with charge density σ is $E = \dfrac{\sigma}{2\epsilon_0}$. Thus,

$$\nu = \frac{1}{2\pi}\sqrt{\frac{q\left(\dfrac{\sigma}{2\epsilon_0}\right)}{m\ell}} = \frac{1}{2\pi}\sqrt{\frac{q\sigma}{2\epsilon_0 m\ell}}.$$

16.61

a)   The force on the electron is

$$\vec{\mathbf{F}} = q\vec{\mathbf{E}} = -e\vec{\mathbf{E}}.$$

To slow the electron, $\vec{\mathbf{F}}$ should be antiparallel to $\vec{\mathbf{v}}$, so $\vec{\mathbf{E}}$ must be parallel to $\vec{\mathbf{v}}$.

b)   Choose a coordinate system with $\hat{\mathbf{i}}$ parallel to $\vec{\mathbf{v}}$ and origin at the place where the electron enters the field.

c)   Apply Newton's second law to the electron in the x direction in order to find a_x.

$$F_x = ma_x \implies a_x = \frac{F_x}{m} = \frac{-eE}{m} = \frac{(-1.602 \times 10^{-19} \text{ C})(100 \text{ N/C})}{9.11 \times 10^{-31} \text{ kg}} = -1.76 \times 10^{13} \text{ m/s}^2\,.$$

d)   Use the velocity component equation for motion with a constant acceleration,

$$v_x(t) = v_{x0} + a_x t.$$

So, when the electron is at rest, we have

$$0 \text{ m/s} = v_{x0} + a_x t \implies t = \frac{-v_{x0}}{a_x} = \frac{-5.00 \times 10^6 \text{ m/s}}{-1.76 \times 10^{13} \text{ m/s}^2} = 2.84 \times 10^{-7} \text{ s}\,.$$

e)   Use the above expressions for a_x and t in the position equation for motion with a constant acceleration,

$$x(t) = x_0 + v_{x0}t + a_x \frac{t^2}{2}$$

$$= 0 \text{ m} + (5.00 \times 10^6 \text{ m/s})(2.84 \times 10^{-7} \text{ s}) + (-1.76 \times 10^{13} \text{ m/s}^2)\frac{(2.84 \times 10^{-7} \text{ s})^2}{2} = 0.71 \text{ m}\,.$$

f)   The electron will not remain at rest because the electric force still is acting on the electron. The electron will accelerate back in the direction from which it came, and emerge with the same speed with which it entered the field.

16.65   Choose a coordinate system with $\hat{\mathbf{i}}$ pointing in the direction of travel of the electron, and with origin at the point where the electron enters the field. Then, in this coordinate system, the electric force on the electron is

$$\vec{\mathbf{F}} = q\vec{\mathbf{E}} = qE\hat{\mathbf{i}} = -eE\hat{\mathbf{i}},$$

so, applying Newton's second law, the acceleration $\vec{\mathbf{a}}$ satisfies

$$\vec{\mathbf{F}} = m\vec{\mathbf{a}} \implies -eE\hat{\mathbf{i}} = ma_x\hat{\mathbf{i}} \implies a_x = \frac{-eE}{m},$$

where $m = 9.11 \times 10^{-31}$ kg is the mass of the electron.

From the kinematic equation for motion with a constant acceleration, the electron's velocity component at any time t is $v_x(t) = v_{x0} + a_x t$. Thus, at the time t that the velocity component is zero,

$$0 \text{ m/s} = v_{x0} + a_x t \implies t = \frac{-v_{x0}}{a_x} = \frac{-v_{x0}}{\dfrac{-eE}{m}} = \frac{mv_{x0}}{eE}.$$

Use this time t and the above expression for a_x in the position kinematic equation for motion with a constant acceleration to find the distance the electron travels.

$$x(t) = x_0 + v_{x0}t + a_x \frac{t^2}{2}$$

$$= 0 \text{ m} + v_{x0}\left(\frac{mv_{x0}}{eE}\right) + \left(\frac{-eE}{m}\right)\frac{\left(\dfrac{mv_{x0}}{eE}\right)^2}{2}$$

$$= \frac{mv_{x0}^2}{2eE}$$

$$= \frac{(9.11 \times 10^{-31} \text{ kg})(5.00 \times 10^6 \text{ m/s})^2}{2(1.602 \times 10^{-19} \text{ C})(500 \text{ N/C})} = 0.142 \text{ m}\,.$$

16.69 Gauss's law for the electric field states that the flux of the electric field through a closed surface is equal to the net (that is, total) charge enclosed by that surface divided by ϵ_0. An electric dipole consists of two charges $+q$ and $-q$ that are equal in magnitude but have opposite sign. Hence, the total charge is 0 C. Since the surface encloses the electric dipole, the flux of the electric field through the surface is also zero.

16.73 There is no charge inside the spitoon, so one might think that, in accordance with Gauss's law, the total flux through it is zero. BUT the spitoon is not a closed surface. Its top is open (otherwise it would not work very well as a spitoon!)

We can still use Gauss's law to advantage, however. Make a lid for the spitoon. The lid is a simple disk with radius 6 cm. When placed carefully on top of the spitoon, it turns it into a closed surface. Therefore we may use Gauss's law on the spittoon-with-lid:

$$\int_{\text{spittoon}} \vec{E} \bullet d\vec{S} + \int_{\text{lid}} \vec{E} \bullet d\vec{S} = \int_{\text{spittoon-with-lid}} \vec{E} \bullet d\vec{S} = 0 \text{ N·m}^2/\text{C}$$

$$\Longrightarrow \int_{\text{spittoon}} \vec{E} \bullet d\vec{S} = -\int_{\text{lid}} \vec{E} \bullet d\vec{S}$$

$$\Longrightarrow \Phi_{\text{spittoon}} = -\Phi_{\text{lid}}.$$

So we need only compute the flux through the lid. Everywhere on the lid, the differential area vector $d\vec{S}$ is parallel to $\vec{E}$, so on the lid, $\vec{E} \bullet d\vec{S} = E\, dA$, the differential flux is just the magnitude E of the electric field multiplied by the differential area dA. Since E is constant, it may be brought outside the integral. Hence

$$\Phi_{\text{lid}} = \int_{\text{lid}} \vec{E} \bullet d\vec{S} = \int_{\text{lid}} E\, dA = E \int_{\text{lid}} dA = E\,(\text{area of lid}) = (200 \text{ N/C})[\pi(6.00 \times 10^{-2} \text{ m})^2] = 2.3 \text{ N·m}^2/\text{C}.$$

Therefore

$$\Phi_{\text{spittoon}} = -\Phi_{\text{lid}} = -2.3 \text{ N·m}^2/\text{C}.$$

Notice, however, that since the surface is not a closed surface, it has (strictly speaking) no "inside," so there is ambiguity in the direction to orient the differential vectors $d\vec{S}$. Had we made the other choice of orientation, our answer would have been $\Phi_{\text{spittoon}} = +2.3 \text{ N·m}^2/\text{C}$. One way of "closing up" the spittoon so that what appeared above to be the outside becomes the inside is as follows: Make a cylindrical can taller than the spittoon and wide enough to contain it. Carefully lower the spittoon into the can so that its top is just level with the top of the spittoon. While you are holding it there, have friend put a lid on the can with a 12 cm diameter hole in it. Weld the top edge of the spittoon to the edge of the hole in the lid. There! Now you have a closed surface whose inside contains what we used to think of as the outside of the spittoon. If we use this surface to compute Φ_{spittoon}, we end up with

$$\Phi_{\text{spittoon}} = +2.3 \text{ N·m}^2/\text{C}.$$

Finally, for those students who have had some vector calculus, notice that the vector field $\vec{E} = E\hat{\mathbf{k}}$ is the curl of the vector field $Ex\hat{\mathbf{j}}$. So by Stokes' theorem, the integral of $\vec{E}$ over the entire spittoon is the integral of $Ex\hat{\mathbf{j}}$ around the boundary of the spittoon. The boundary is a circle of radius 6 cm that we may take to be in the x-y-plane centered at the origin, so the integral of $Ex\hat{\mathbf{j}}$ around the boundary is fairly easy to compute. When we do, we again find $\pm E\pi(6.00 \times 10^{-2} \text{ m})^2$. Here the ambiguity of sign is due to choice about which to go around the circle.

16.77

a)   Here's the sketch.

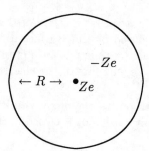

b)   Gauss's law states that flux of the electric field over a closed surface is equal to the net charge enclosed by that surface divided by ϵ_0:

(1)
$$\int_{\text{closed surface}} \vec{\mathbf{E}} \bullet d\vec{\mathbf{S}} = \frac{q_{\text{enclosed}}}{\epsilon_0}.$$

Use a concentric sphere whose radius r is greater than R as a Gaussian surface. By symmetry, the magnitude of the electric field is constant over the entire Gaussian surface and points radially outward, parallel to $d\vec{\mathbf{S}}$ at every location on the surface. Hence the left-hand side of (1) can be evaluated explicitly:

$$4\pi r^2 E = \frac{q_{\text{enclosed}}}{\epsilon_0}.$$

However, the total charge enclosed by this Gaussian surface is 0 C. Therefore:

$$4\pi r^2 E = \frac{0 \text{ C}}{\epsilon_0} = 0 \text{ N·m}^2/\text{C} \implies E = 0 \text{ N/C}.$$

c)   Use a concentric sphere whose radius r is less than R as a Gaussian surface. By symmetry, the magnitude of the electric field is constant over the entire Gaussian surface and points radially outward, parallel to $d\vec{\mathbf{S}}$ at every location on the surface. Hence the left-hand side of Gauss's law can again be evaluated explicitly:

(1)
$$4\pi r^2 E = \frac{q_{\text{enclosed}}}{\epsilon_0}.$$

The total charge enclosed by this Gaussian surface is the nuclear charge Ze plus that portion of the charge $-Ze$ that lies within the Gaussian surface. Since the charge $-Ze$ is distributed uniformly throughout the volume, the portion within the Gaussian sphere of radius r is the ratio of the volume of the sphere of radius r to the volume of the sphere of radius R. Hence the charge enclosed by the Gaussian surface is

$$q_{\text{enclosed}} = Ze + \frac{\left(\dfrac{4\pi r^3}{3}\right)}{\left(\dfrac{4\pi R^3}{3}\right)}(-Ze) = Ze\left(1 - \frac{r^3}{R^3}\right).$$

Hence, from equation (1),

$$E = \frac{Ze\left(1 - \dfrac{r^3}{R^3}\right)}{4\pi\epsilon_0 \, r^2}.$$

Note that the magnitude of this electric field goes continuously to zero as $r \to R$, until at $r = R$, we have $E_{\text{inside}} = 0 \text{ N/C} = E_{\text{outside}}$.

Chapter 17

Electric Potential Energy and the Electric Potential

17.1   The work done by the electrical force is

$$W_{elec} = -\Delta PE = -(PE_f - PE_i) = -(qV_f - qV_i)$$
$$= -[(-5.00 \times 10^{-6} \text{ C})(150 \text{ V}) - (-5.00 \times 10^{-6} \text{ C})(0 \text{ V})] = 7.50 \times 10^{-4} \text{ J}.$$

17.5

a)   As in Strategic Example 17.1 on page 770 of the text, we model the bottom of the thundercloud and the surface of the Earth as two parallel charged plates with uniform charge densities of σ and $-\sigma$. For this geometry the magnitude of the potential difference between the two plates is $|\Delta V| = Ed$ where E is the magnitude of the electric field and d is the distance between the surface of the Earth and the bottom of the cloud. Thus,

$$|\Delta V| = Ed = (3.0 \times 10^6 \text{ V/m})(1.00 \times 10^3 \text{ m}) = 3.0 \times 10^9 \text{ V}.$$

b)   From Example 16.17 on page 738 of the text, the magnitude E of the electric field between the two parallel plates is $E = \dfrac{\sigma}{\epsilon_0}$. Therefore,

$$\sigma = \epsilon_0 E = \left(\frac{1}{4\pi}\right) \frac{1}{\left(\dfrac{1}{4\pi\epsilon_0}\right)} E = \left(\frac{1}{4\pi}\right) \frac{1}{9.00 \times 10^9 \text{ N·m}^2/\text{C}^2} (3.0 \times 10^9 \text{ V}) = 2.7 \times 10^{-5} \text{ C/m}^2.$$

The total charge on the cloud is the surface charge density times the area of the bottom of the cloud.

$$Q = (2.7 \times 10^{-5} \text{ C/m}^2)(2.0 \text{ km}^2) = (2.7 \times 10^{-5} \text{ C/m}^2)(2.0 \times 10^6 \text{ m}^2) = 54 \text{ C}.$$

17.9

a)   The electrical potential is the potential energy per unit of charge. Its SI units are joules per coulomb. By analogy, the gravitational potential is the potential energy per unit mass. Its SI units are joules per kilogram.

b)   The gravitational potential difference is the difference in the gravitational potential at the two locations. Let $\hat{\mathbf{j}}$ point up, and choose the origin at ground level. Then the difference in gravitational potential is

$$\frac{\mathrm{PE_f}}{m} - \frac{\mathrm{PE_i}}{m} = \frac{mgy_\mathrm{f}}{m} - \frac{mgy_\mathrm{i}}{m} = g(y_\mathrm{f} - y_\mathrm{i}) = (9.81 \text{ m/s}^2)(8.00 \text{ m} - 3.00 \text{ m}) = 49.1 \text{ J/kg}.$$

17.13

a)   From Example 16.17 on page 738 of the text, the magnitude E of the electric field between two parallel plates is related to the magnitude of the surface charge density σ on each by

$$E = \frac{\sigma}{\epsilon_0} \implies$$

$$\sigma = \epsilon_0 E = \left(\frac{1}{4\pi}\right)\frac{1}{\left(\frac{1}{4\pi\epsilon_0}\right)}E = \left(\frac{1}{4\pi}\right)\left(\frac{1}{9.00 \times 10^9 \text{ N·m}^2/\text{C}^2}\right)(1.00 \times 10^4 \text{ N/C}) = 8.84 \times 10^{-8} \text{ C/m}^2.$$

Since electric field lines go from positive to negative charge, the top plate has the negative charge. Thus,

$$\sigma_{\text{top plate}} = -8.84 \times 10^{-8} \text{ C/m}^2.$$

b)   The electric field lines go from positive to negative charge, so the bottom plate has positive charge. Thus,

$$\sigma_{\text{bottom plate}} = 8.84 \times 10^{-8} \text{ C/m}^2.$$

c)   The absolute value of the potential difference between the parallel plates is related to the field and separation by

$$|\Delta V| = Ed = (1.00 \times 10^4 \text{ N/C})(0.100 \text{ m}) = 1.00 \times 10^3 \text{ V}.$$

Electric field lines go from regions of high potential to regions of low potential. If the lower plate is "grounded," then by convention $V_{\text{bottom}} = 0$ V. The upper plate has a lower potential, so $V_{\text{top plate}} = -1.00 \times 10^3$ V.

d)   Choose a coordinate system with $\hat{\mathbf{j}}$ pointing up. Let $m = 1.00 \times 10^{-5}$ kg denote the mass of the particle. Then the two forces on the particle are:

1. its weight $\vec{\mathbf{w}} = -mg\hat{\mathbf{j}}$; and

2. the electrical force $\vec{\mathbf{F}}_{\text{elec}} = qE\hat{\mathbf{j}}$.

Since the particle is not accelerating, the sum of these forces is zero, so

$$-mg\hat{\mathbf{j}} + qE\hat{\mathbf{j}} = \mathbf{0} \text{ N} \implies q = \frac{mg}{E} = \frac{(1.00 \times 10^{-5} \text{ kg})(9.81 \text{ m/s}^2)}{1.00 \times 10^4 \text{ N/C}} = 9.81 \times 10^{-9} \text{ C}.$$

17.17   Let $\hat{\mathbf{j}}$ point up, and choose the origin at ground level. Then the gravitational potential energy of a mass m is $\mathrm{PE}_{\text{gravity}} = mgy$. The gravitational potential V_{gravity} is the gravitational potential energy per unit mass, so

$$V_{\text{gravity}} = \frac{\mathrm{PE}_{\text{gravity}}}{m} = \frac{mgy}{m} = gy.$$

17.21 The magnitude of the electric field on the surface of a spherical conductor of radius R is

(1) $$E = \frac{1}{4\pi\epsilon_0}\frac{Q}{R^2}.$$

The electric potential V on the surface of such a sphere is that of a point like charge with r equal to the radius R of the sphere.

$$V = \frac{1}{4\pi\epsilon_0}\frac{Q}{R}.$$

Hence,

$$E = \frac{V}{R}.$$

We want $E < E_{\max} = 3.0 \times 10^6$ V/m. Therefore

$$\frac{V}{R} < E_{\max} \implies R > \frac{V}{E_{\max}} = \frac{25.0 \times 10^3 \text{ V}}{3.0 \times 10^6 \text{ V/m}} = 8.3 \times 10^{-3} \text{ m}.$$

If the radius is smaller, the magnitude of the field increases (see equation (1) above), so this radius is a minimum radius if you do not want E to exceed 3.0×10^6 V/m on the surface of the sphere.

17.25

a) For the region $r > R$:
The nucleus with charge $+Ze$ acts as a point-like charge and produces a potential

$$V_{\text{nucleus}} = \frac{1}{4\pi\epsilon_0}\left(\frac{Ze}{r}\right).$$

For the region outside the atom, the uniform spherical volume of charge $-Ze$ of the nonnucleus acts the same as a point-like charge located at its center and so produces a potential

$$V_{\text{nonnucleus}} = \frac{1}{4\pi\epsilon_0}\left(\frac{-Ze}{r}\right).$$

The total potential is

$$V_{\text{outside}} = V_{\text{nucleus}} + V_{\text{nonnucleus}} = \frac{1}{4\pi\epsilon_0}\left(\frac{Ze}{r}\right) + \frac{1}{4\pi\epsilon_0}\left(\frac{-Ze}{r}\right) = 0 \text{ V}.$$

b) For the region $r < R$:
The nucleus with charge $+Ze$ is still a point-like charge and produces a potential

$$V_{\text{nucleus}} = \frac{1}{4\pi\epsilon_0}\left(\frac{Ze}{r}\right).$$

From Example 17.8 on pages 777-778 of the text, the spherical volume of charge $-Ze$ produces a potential within it of

$$V_{\text{nonnucleus}} = \frac{1}{4\pi\epsilon_0}\left(\frac{-Ze}{2R}\left(3 - \frac{r^2}{R^2}\right)\right).$$

Hence,

$$V_{\text{inside}} = V_{\text{nucleus}} + V_{\text{nonnucleus}} = \frac{1}{4\pi\epsilon_0}\left(\frac{Ze}{r}\right) + \frac{1}{4\pi\epsilon_0}\left(\frac{-Ze}{2R}\left(3 - \frac{r^2}{R^2}\right)\right) = \frac{Ze}{4\pi\epsilon_0}\left(\frac{1}{r} - \frac{3}{2R} + \frac{r^2}{2R^3}\right).$$

Notice that when $r = R$, we get 0 V just as in part a). Thus, both potentials have the surface of the atom as a common ground.

17.29

a)   The electric potential on the surface of the sphere is

$$V = \frac{1}{4\pi\epsilon_0}\frac{Q}{R} = (9.00 \times 10^9 \text{ N·m}^2/\text{C}^2)\left(\frac{6.00 \times 10^{-9} \text{ C}}{5.00 \times 10^{-2} \text{ m}}\right) = 1.08 \times 10^3 \text{ V}.$$

b)   From Example 17.8 on pages 777-778 of the text, the electric potential at the center of the sphere where $r = 0$ m is

$$V = \frac{1}{4\pi\epsilon_0}\frac{3Q}{2R} = (9.00 \times 10^9 \text{ N·m}^2/\text{C}^2)\frac{3(6.00 \times 10^{-9} \text{ C})}{2(5.00 \times 10^{-2} \text{ m})} = 1.62 \times 10^3 \text{ V}.$$

17.33

a)   The electric potential on the surface of a uniformly distributed spherical charge is

$$V = \frac{1}{4\pi\epsilon_0}\frac{Q}{R}$$

where Q is the total charge and R is the radius of the sphere. Hence, for the 2.00 cm sphere,

$$V_{2.00 \text{ cm}} = (9.00 \times 10^9 \text{ N·m}^2/\text{C}^2)\left(\frac{3.00 \times 10^{-9} \text{ C}}{2.00 \times 10^{-2} \text{ m}}\right) = 1.35 \times 10^3 \text{ V}.$$

For the surface of the 4.00 cm sphere,

$$V_{4.00 \text{ cm}} = (9.00 \times 10^9 \text{ N·m}^2/\text{C}^2)\left(\frac{3.00 \times 10^{-9} \text{ C}}{4.00 \times 10^{-2} \text{ m}}\right) = 675 \text{ V}.$$

b)   When the two are connected by a conducting wire their potentials equalize, so

$$\frac{1}{4\pi\epsilon_0}\left(\frac{Q_{2.00 \text{ cm}}}{2.00 \times 10^{-2} \text{ m}}\right) = \frac{1}{4\pi\epsilon_0}\left(\frac{Q_{4.00 \text{ cm}}}{4.00 \times 10^{-2} \text{ m}}\right) \implies Q_{4.00 \text{ cm}} = 2.00 Q_{2.00 \text{ cm}}.$$

Since charge is conserved, the total charge remains the same, hence

$$Q_{2.00 \text{ cm}} + Q_{4.00 \text{ cm}} = 3.00 \times 10^{-9} \text{ C} + 3.00 \times 10^{-9} \text{ C} = 6.00 \times 10^{-9} \text{ C}.$$
$$\implies 3Q_{2.00 \text{ cm}} = 6.00 \times 10^{-9} \text{ C}$$
$$\implies Q_{2.00 \text{ cm}} = 2.00 \times 10^{-9} \text{ C} \quad \text{and} \quad Q_{4.00 \text{ cm}} = 4.00 \times 10^{-9} \text{ C}.$$

The potential on the surface of the 2.00 cm sphere is now

$$V_{2.00 \text{ cm}} = \frac{1}{4\pi\epsilon_0}\left(\frac{Q_{2.00 \text{ cm}}}{2.00 \times 10^{-2} \text{ m}}\right) = (9.00 \times 10^9 \text{ N·m}^2/\text{C}^2)\left(\frac{2.00 \times 10^{-9} \text{ C}}{2.00 \times 10^{-2} \text{ m}}\right) = 900 \text{ V}.$$

The potential on the surface of the 4.00 cm is, of course, now the same at

$$V_{4.00 \text{ cm}} = \frac{1}{4\pi\epsilon_0}\left(\frac{Q_{4.00 \text{ cm}}}{4.00 \times 10^{-2} \text{ m}}\right) = (9.00 \times 10^9 \text{ N·m}^2/\text{C}^2)\left(\frac{4.00 \times 10^{-9} \text{ C}}{4.00 \times 10^{-2} \text{ m}}\right) = 900 \text{ V}.$$

The original charge on each was 3.00×10^{-9} C, and the new charges are 2.00×10^{-9} C and 4.00×10^{-9} C, so 1.00×10^{-9} C was exchanged in the process.

17.37

a) Let P_r be a point a distance r from the charged line. Choose the origin of a coordinate system to be the point on the line closest to P. Let $\hat{\mathbf{r}}$ be a unit vector pointing from the origin towards P_r, so then $r\hat{\mathbf{r}}$ is the position vector for P_r. Let P_s be the point with position vector $s\hat{\mathbf{r}}$, so P_a is a distance a from the line, and, in general, if s is between a and r, then P_s is on the straight line segment joining P_a to P_r and is a distance s from the charged line. Here's the picture.

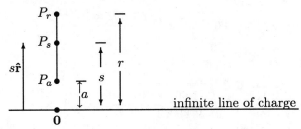

From Table 16.2 on page 734 of the text, the field at P_s is $\vec{\mathbf{E}} = \dfrac{1}{4\pi\epsilon_0}\dfrac{2\lambda}{s}\hat{\mathbf{r}}$. The differential change $d\vec{\mathbf{r}}$ along the path from P_a to P_r is $d\vec{\mathbf{r}} = ds\,\hat{\mathbf{r}}$. Therefore

$$
\begin{aligned}
V(r) - V(a) &= -\int_{a\hat{\mathbf{r}}}^{r\hat{\mathbf{r}}} \vec{\mathbf{E}} \bullet d\vec{\mathbf{r}} \\
&= -\int_{a}^{r} \frac{1}{4\pi\epsilon_0}\frac{2\lambda}{s}\hat{\mathbf{r}} \bullet ds\,\hat{\mathbf{r}} \\
&= -\int_{a}^{r} \frac{1}{4\pi\epsilon_0}\frac{2\lambda}{s}ds \\
&= -\frac{\lambda}{2\pi\epsilon_0}\ln s\Big|_{a}^{r} \\
&= -\frac{\lambda}{2\pi\epsilon_0}(\ln r - \ln a) = \frac{\lambda}{2\pi\epsilon_0}\ln\left(\frac{a}{r}\right).
\end{aligned}
$$

b) and

c)

$$
V(r) > 0\text{ V} \iff \frac{\lambda}{2\pi\epsilon_0}\ln\left(\frac{a}{r}\right) > 0\text{ V} \iff \ln\left(\frac{a}{r}\right) > 0 \iff \frac{a}{r} > 1 \iff r < a.
$$

Similarly,

$$
V(r) < 0\text{ V} \iff r > a.
$$

d) Since $V(r) - V(a) = \dfrac{\lambda}{2\pi\epsilon_0}(\ln a - \ln r)$, and neither $\ln 0$ nor $\lim_{r\to\infty}\ln r$ are defined (as finite numbers), we cannot choose electrical ground to be at $r = 0$ m, nor can we choose it at $r = \infty$ m.

e)

$$
-\frac{d}{dr}V(r) = -\frac{d}{dr}\frac{\lambda}{2\pi\epsilon_0}(\ln a - \ln r) = \frac{\lambda}{2\pi\epsilon_0}\frac{1}{r} = E_r.
$$

17.41

a) The absolute value of the change in the kinetic energy of an electron expressed in electron-volts is numerically the same as the absolute value of the potential difference, expressed in volts, through which the particle was accelerated. Hence, the change in the kinetic energy of the electron is

$$
|\Delta\text{KE}| = 1.00 \times 10^3 \text{ eV}.
$$

b) Let v be the final speed of the electron. Since the electron began at rest, $\Delta KE = \frac{1}{2}mv^2$, so

$$(1) \qquad\qquad v = \sqrt{\frac{2\Delta KE}{m}}.$$

Convert the value of ΔKE found in part a) from electron volts to joules.

$$\Delta KE = 1.00 \times 10^3 \text{ eV} = (1.00 \times 10^3 \text{ eV})\left(\frac{1.602 \times 10^{-19} \text{ J}}{\text{eV}}\right) = 1.602 \times 10^{-16} \text{ J}.$$

Now use this in equation (1).

$$v = \sqrt{\frac{2\Delta KE}{m}} = \sqrt{\frac{2(1.602 \times 10^{-16} \text{ J})}{9.11 \times 10^{-31} \text{ kg}}} = 1.88 \times 10^7 \text{ m/s}.$$

17.45

a) The electric field is directed from regions of higher to regions of lower electric potential, so it is directed from the 0 V sheet to the -200 V sheet. Since the positron has positive charge, the electric force on it is parallel to the field ($\vec{F} = q\vec{E}$). Therefore, release the positron near the 0 V sheet.

b) The change in the kinetic energy of the positron expressed in electron-volts is numerically the same as the potential difference, expressed in volts, through which the particle was accelerated. Hence, the change in the kinetic energy of the positron is

$$\Delta KE = 200 \text{ eV}.$$

Converted to joules, this is

$$\Delta KE = 200 \text{ eV} = (200 \text{ eV})\left(\frac{1.602 \times 10^{-19} \text{ J}}{\text{eV}}\right) = 3.20 \times 10^{-17} \text{ J}.$$

c) Let v be the final speed of the positron. It was released at rest, so

$$\Delta KE = \frac{1}{2}mv^2 - 0 \text{ J} \implies v = \sqrt{\frac{2\Delta KE}{m}} = \sqrt{\frac{2(3.20 \times 10^{-17} \text{ J})}{9.11 \times 10^{-31} \text{ kg}}} = 8.38 \times 10^6 \text{ m/s}.$$

17.49

a) The electric field is directed from regions of higher to regions of lower electric potential, so the electric field is directed from the 1.00 kV to the 0 V sheet.

b) The force on the electron is

$$\vec{F} = q\vec{E} = (-e)\vec{E} = -e\vec{E}.$$

Hence, the direction of the force on the electron is *opposite* the direction of the field.

c) The total mechanical energy at point A is

$$E_A = KE_A + PE_A = \frac{1}{2}mv_A^2 + qV_A$$

$$= \frac{1}{2}(9.11 \times 10^{-31} \text{ kg})(5.0 \times 10^7 \text{ m/s})^2 + (-1.602 \times 10^{-19} \text{ C})(1.00 \times 10^3 \text{ V})$$

$$= 1.1 \times 10^{-15} \text{ J} - 1.6 \times 10^{-16} \text{ J} = 0.9 \times 10^{-15} \text{ J}.$$

Some students may carry one or more extra digits in computing this, and then round to one significant digit at the end. This will result in the equally acceptable (arguably more acceptable) answer of 0.1×10^{-14} J.

d)   Since the conservative electric force is the only force on the electron, the total mechanical energy of the electron is conserved. Therefore, the total mechanical energy at point B is the same as it is at point A.

$$E_B = E_A = 0.9 \times 10^{-15} \text{ J}.$$

e)   At point B the total mechanical energy is

$$E_B = \text{KE}_B + \text{PE}_B = \frac{1}{2}mv_B^2 + qV_B$$

$$\implies v_B = \sqrt{\frac{2(E_B - qV_B)}{m}} = \sqrt{\frac{2[(0.9 \times 10^{-15} \text{ J}) - (-1.602 \times 10^{-19} \text{ C})(0 \text{ V})]}{9.11 \times 10^{-31} \text{ kg}}} = 4 \times 10^7 \text{ m/s}.$$

17.53

a)   The change in the kinetic energy is

$$\Delta\text{KE} = -q\Delta V = -2e(20 \text{ V}) = -40 \text{ eV}.$$

b)   The change in the kinetic energy is

$$\Delta\text{KE} = -q\Delta V = -2e(-30 \text{ V}) = 60 \text{ eV}.$$

17.57

a)   The potential on the surface of the sphere is

$$V = \frac{1}{4\pi\epsilon_0}\frac{Q}{R}.$$

b)   The potential energy of the charge is

$$\text{PE} = -|q|V = -\frac{1}{4\pi\epsilon_0}\frac{|q|Q}{R}.$$

c)   The only force acting on the escaping charge is the conservative electrical force. Apply the CWE theorem:

$$0 \text{ J} = W_{\text{nonconservative}} = \Delta(\text{KE} + \text{PE})$$

$$\implies 0 \text{ J} = (\text{KE}_f + \text{PE}_f) - (\text{KE}_i + \text{PE}_i) = (0 \text{ J} + 0 \text{ J}) - \left(\frac{1}{2}mv^2 + \left(-\frac{1}{4\pi\epsilon_0}\frac{|q|Q}{R}\right)\right)$$

$$\implies v = \sqrt{\frac{2|q|Q}{4\pi\epsilon_0 Rm}}.$$

d)   The gravitational potential energy is proportional to the mass, and so appears in both the KE and PE terms in the CWE theorem and therefore factors out of their sum. The electrical potential energy is independent of the mass, so the mass appears in the KE term, but not the PE term and so does not factor out of their sum. Rather, v depends on the charge to mass ratio $\dfrac{|q|}{m}$.

e)   When we substitute $v = c$ in the result of part c) and solve for Q we have

$$Q = \frac{4\pi\epsilon_0 Rmc^2}{2|q|} = \frac{1}{\left(\dfrac{1}{4\pi\epsilon_0}\right)}\frac{Rmc^2}{2|q|}$$

$$= \frac{1}{9.00 \times 10^9 \text{ N·m}^2/\text{C}^2}\frac{(1 \times 10^{-14} \text{ m})(9.11 \times 10^{-31} \text{ kg})(3.00 \times 10^8 \text{ m/s})^2}{2(1.602 \times 10^{-19} \text{ C})} = 3 \times 10^{-19} \text{ C}.$$

This is about the charge of 2 protons ($\approx 3.20 \times 10^{-19}$ C).

f)   If the electron is placed 1×10^{-10} m from the nuclear charge, then

$$Q = \frac{4\pi\epsilon_0 Rmc^2}{2|q|} = \frac{1}{\left(\dfrac{1}{4\pi\epsilon_0}\right)} \frac{Rmc^2}{2|q|}$$

$$= \frac{1}{9.00 \times 10^9 \ \text{N·m}^2/\text{C}^2} \frac{(1 \times 10^{-10} \ \text{m})(9.11 \times 10^{-31} \ \text{kg})(3.00 \times 10^8 \ \text{m/s})^2}{2(1.602 \times 10^{-19} \ \text{C})} = 3 \times 10^{-15} \ \text{C}.$$

This is about the charge of 2×10^4 protons. Nuclei with such large charge quantum numbers do not exist in nature, so such electrical black holes do not exist.

17.61

a)   The total electric field $\vec{E}$ is the vector sum $\vec{E} = \vec{E}_1 + \vec{E}_2$ of two fields: the electric field $\vec{E}_1$ produced by the -3.00×10^{-6} C charge, and the electric field $\vec{E}_2$ produced by the 2.00×10^{-6} C charge. We'll compute each of these at the point P and then add them together (as vectors!).

The distance from P to the -3.00×10^{-6} C charge is $r_1 = 5.00 \times 10^{-2}$ m. Hence, the magnitude of $\vec{E}_1$ at P is

$$E_1 = \frac{1}{4\pi\epsilon_0} \frac{|Q_1|}{r_1^2} = (9.00 \times 10^9 \ \text{N·m}^2/\text{C}^2) \frac{3.00 \times 10^{-6} \ \text{C}}{(5.00 \times 10^{-2} \ \text{m})^2} = 1.08 \times 10^7 \ \text{N/C}.$$

The charge is negative, so the field points from P towards the charge. The unit vector $\hat{\mathbf{r}}_1$ in that direction is $\hat{\mathbf{r}}_1 = -\hat{\mathbf{j}}$, so

$$(1) \qquad\qquad \vec{E}_1 = E_1\hat{\mathbf{r}}_1 = E_1(-\hat{\mathbf{j}}) = -(1.08 \times 10^7 \ \text{N/C})\hat{\mathbf{j}}.$$

The 2.00×10^{-6} C charge is $\sqrt{(0.1200 \ \text{m})^2 + (0.0500 \ \text{m})^2} = 0.1300$ m from P. Hence, E_2 at P is

$$E_2 = \frac{1}{4\pi\epsilon_0} \frac{|Q_2|}{r_2^2} = (9.00 \times 10^9 \ \text{N·m}^2/\text{C}^2) \frac{2.00 \times 10^{-6} \ \text{C}}{(0.1300 \ \text{m})^2} = 1.07 \times 10^6 \ \text{N/C}.$$

The charge is positive, so the field at P points away from the charge. The unit vector in that direction is

$$\hat{\mathbf{r}}_2 = \frac{1}{0.1300 \ \text{m}} \left((-0.1200 \ \text{m})\hat{\mathbf{i}} + (0.0500 \ \text{m})\hat{\mathbf{j}}\right) = (-0.923)\hat{\mathbf{i}} + (0.385)\hat{\mathbf{j}}.$$

Thus,

$$(2) \quad \vec{E}_2 = E_2\hat{\mathbf{r}}_2 = (1.07 \times 10^6 \ \text{N/C})\left(-0.923\hat{\mathbf{i}} + 0.385\hat{\mathbf{j}}\right) = (-0.988 \times 10^6 \ \text{N/C})\hat{\mathbf{i}} + (0.412 \times 10^6 \ \text{N/C})\hat{\mathbf{j}}.$$

Adding (1) and (2) we have

$$\vec{E} = \vec{E}_1 + \vec{E}_2 = -(0.988 \times 10^6 \ \text{N/C})\hat{\mathbf{i}} - (1.04 \times 10^7 \ \text{N/C})\hat{\mathbf{j}} = -(0.01 \times 10^7 \ \text{N/C})\hat{\mathbf{i}} - (1.04 \times 10^7 \ \text{N/C})\hat{\mathbf{j}}.$$

b)   The electric potential at P is the scalar sum $V = V_1 + V_2$ of the potential V_1 produced by the -3.00×10^{-6} C charge and the potential V_2 produced by the 2.00×10^{-6} C charge:

$$V_1 = \frac{1}{4\pi\epsilon_0} \frac{Q_1}{r_1} = (9.00 \times 10^9 \ \text{N·m}^2/\text{C}^2) \left(\frac{-3.00 \times 10^{-6} \ \text{C}}{0.0500 \ \text{m}}\right) = -5.40 \times 10^5 \ \text{V},$$

and

$$V_2 = \frac{1}{4\pi\epsilon_0} \frac{Q_2}{r_2} = (9.00 \times 10^9 \ \text{N·m}^2/\text{C}^2) \left(\frac{2.00 \times 10^{-6} \ \text{C}}{0.1300 \ \text{m}}\right) = 1.38 \times 10^5 \ \text{V},$$

so

$$V = V_1 + V_2 = -5.40 \times 10^5 \ \text{V} + 1.38 \times 10^5 \ \text{V} = -4.02 \times 10^5 \ \text{V}.$$

c) The potential energy of the dipole is

$$\text{PE} = -\vec{\mathbf{p}} \bullet \vec{\mathbf{E}} = -(6.0 \times 10^{-30} \text{ C·m})\hat{\mathbf{j}} \bullet \left(-(0.01 \times 10^7 \text{ N/C})\hat{\mathbf{i}} - (1.04 \times 10^7 \text{ N/C})\hat{\mathbf{j}}\right) = 6.2 \times 10^{-23} \text{ J}.$$

d) The torque on the dipole is

$$\vec{\tau} = \vec{\mathbf{p}} \times \vec{\mathbf{E}} = (6.0 \times 10^{-30} \text{ C·m})\hat{\mathbf{j}} \times \left(-(0.01 \times 10^7 \text{ N/C})\hat{\mathbf{i}} - (1.04 \times 10^7 \text{ N/C})\hat{\mathbf{j}}\right) = (6 \times 10^{-25} \text{ N·m})\hat{\mathbf{k}}.$$

17.65 The potential energy is

$$\text{PE} = \frac{1}{4\pi\epsilon_0}\frac{Q_1 Q_2}{r_{12}} + \frac{1}{4\pi\epsilon_0}\frac{Q_1 Q_3}{r_{13}} + \frac{1}{4\pi\epsilon_0}\frac{Q_2 Q_3}{r_{23}}$$
$$= \frac{1}{4\pi\epsilon_0}\frac{1}{r}\left[\left(\frac{2e}{3}\right)\left(\frac{2e}{3}\right) + \left(\frac{2e}{3}\right)\left(\frac{-e}{3}\right) + \left(\frac{2e}{3}\right)\left(\frac{-e}{3}\right)\right] = 0 \text{ J}.$$

17.69 According to Equation 17.29 on page 788 in the text, the electric potential produced by dipole 1 at any point P is

$$V_1 = \frac{1}{4\pi\epsilon_0}\frac{p_1 \cos\theta}{r_P^2},$$

where θ is the angle that $\vec{\mathbf{p}}_1$ makes with the position vector $\vec{\mathbf{r}}_P$ of P with respect to the position of dipole 1, r_P is the distance from dipole 1 to P (the magnitude of $\vec{\mathbf{r}}_P$), and r_P is large compared to the separation of charges in dipole 1. Now apply this equation to find V_1 when P is the position of the second dipole. Then the angle θ is $0°$ and the distance r_P is r. Hence the potential produced by dipole 1 at the position of dipole 2 is

$$V_1(r) = \frac{1}{4\pi\epsilon_0}\frac{p_1}{r^2},$$

which means that the $\hat{\mathbf{r}}$-component of the electric field $\vec{\mathbf{E}}_1$ produced by the first dipole is

$$E_1 = -\frac{d}{dr}V_1(r) = \frac{1}{4\pi\epsilon_0}\frac{2p_1}{r^3},$$

and the $\hat{\boldsymbol{\theta}}$-component of $\vec{\mathbf{E}}_1$ is zero since P is on the x-axis. Therefore, the electric field at point P is

$$\vec{\mathbf{E}}_1 = \frac{1}{4\pi\epsilon_0}\frac{2p_1}{r^3}\hat{\mathbf{r}},$$

where $\hat{\mathbf{r}}$ is the unit vector pointing from dipole 1 to dipole 2. Since the angle between $\hat{\mathbf{r}}$ and $\vec{\mathbf{p}}_2$ is $0°$, $\vec{\mathbf{p}}_2 \bullet \hat{\mathbf{r}} = p_2|\hat{\mathbf{r}}|\cos 0° = p_2$. Thus, the potential energy of the second dipole in the field $\vec{\mathbf{E}}_1$ is

$$\text{PE} = -\vec{\mathbf{p}}_2 \bullet \vec{\mathbf{E}}_1 = -\vec{\mathbf{p}}_2 \bullet \left(\frac{1}{4\pi\epsilon_0}\frac{2p_1}{r^3}\hat{\mathbf{r}}\right) = -\frac{1}{4\pi\epsilon_0}\frac{2p_1}{r^3}\vec{\mathbf{p}}_2 \bullet \hat{\mathbf{r}} = -\frac{1}{4\pi\epsilon_0}\frac{2p_1}{r^3}p_2 = -\frac{1}{4\pi\epsilon_0}\frac{2p_1 p_2}{r^3}.$$

Chapter 18

Circuit Elements, Independent Voltage Sources, and Capacitors

18.1 Nodes are where two or more circuit elements are connected together. There are three nodes in the circuit. They are depicted below. Nodes 1 and 2 are depicted as single dots. Node 3 contains 3 dots, since they are all at the same potential.

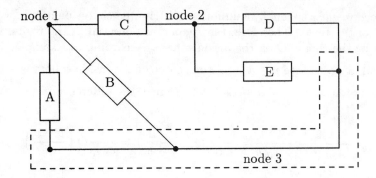

18.5 Circuit elements in series are strung out like beads on a string. There are no circuit elements in Figure P.2 that are in series. Circuit elements in parallel are connected to the same two distinct nodes. Circuit elements A and D are in parallel. Circuit elements B and E also are in parallel.

18.9 Arrange 6 such cells in series, with the positive terminal of one cell connected to the negative terminal to the next, as shown below.

$$A \bullet \!-\!\! \overset{-\quad+}{(2.0\,\text{V})} \!-\! \overset{-\quad+}{(2.0\,\text{V})} \!-\! \overset{-\quad+}{(2.0\,\text{V})} \!-\! \overset{-\quad+}{(2.0\,\text{V})} \!-\! \overset{-\quad+}{(2.0\,\text{V})} \!-\! \overset{-\quad+}{(2.0\,\text{V})} \!-\!\bullet B$$

Since potential differences in series add, the increase in potential from point A to point B is

$$V_{\text{eq}} = 2.0\,\text{V} + 2.0\,\text{V} + 2.0\,\text{V} + 2.0\,\text{V} + 2.0\,\text{V} + 2.0\,\text{V} = 12.0\,\text{V}.$$

So this arrangement of cells is equivalent to the voltage source below.

$$A \bullet \!-\! \overset{-\quad+}{(12.0\,\text{V})} \!-\!\bullet B$$

18.13   By definition of capacitance (Equation 18.1 on page 814 of the text),

$$C = \frac{Q}{V} \implies Q = VC = (1.50 \text{ V})(15 \times 10^{-12} \text{ F}) = 23 \times 10^{-12} \text{ C}.$$

18.17   According to Equation (1) of Example 18.3 on page 815 of the text, the capacitance of a parallel plate capacitor with plate area A and separation distance d is $C = \frac{\epsilon_0 A}{d}$. In this problem, $A = \pi R^2$ and $d = \frac{R}{1000}$, so

$$C = \frac{\epsilon_0 (\pi R^2)}{\left(\frac{R}{1000}\right)} = 1000 \epsilon_0 \pi R.$$

Thus, if R is doubled, then C is also doubled.

18.21   By definition, the capacitance is

$$C = \left| \frac{Q}{V} \right|,$$

Where Q is the charge on one of the conductors, and V is the potential difference between the two conductors. Let Q be the charge on the inner conductor and $-Q$ the charge on the outer conductor. We need to compute V, the potential difference.

The electric potential produced *inside* a sphere by a uniformly distributed charge on its surface is constant (see Example 17.7 on page 777 of the text). Therefore, the *change* in potential in the region between the conductors is produced entirely by the charge Q on the inside sphere. According to Example 17.6 on pages 776-777 of the text, the potential at a distance $r \geq a$ from the center of the inside sphere is $\frac{1}{4\pi\epsilon_0} \frac{Q}{r}$. (This is using the convention that the potential at $r = \infty$ m is 0 V.) Hence, the difference in potential between the two spheres is

$$V = V(a) - V(b) = \frac{1}{4\pi\epsilon_0} \frac{Q}{a} - \frac{1}{4\pi\epsilon_0} \frac{Q}{b} = \frac{1}{4\pi\epsilon_0} Q \left(\frac{1}{a} - \frac{1}{b} \right) = \frac{1}{4\pi\epsilon_0} Q \left(\frac{b-a}{ab} \right).$$

Therefore, the capacitance is

$$C = \left| \frac{Q}{V} \right| = \left| \frac{Q}{\frac{1}{4\pi\epsilon_0} Q \left(\frac{b-a}{ab} \right)} \right| = 4\pi\epsilon_0 \left(\frac{ab}{b-a} \right).$$

18.25   The three capacitors are connected in series, so

$$\frac{1}{C_{\text{eq}}} = \frac{1}{C_1} + \frac{1}{C_2} + \frac{1}{C_3} = \frac{1}{2.0 \ \mu\text{F}} + \frac{1}{3.0 \ \mu\text{F}} + \frac{1}{4.0 \ \mu\text{F}} = \frac{13}{12 \ \mu\text{F}} \implies C_{\text{eq}} = \frac{12}{13} \ \mu\text{F} = 0.92 \ \mu\text{F}.$$

18.29   The 16 μF and 4 μF capacitors are in series. Replace them by an equivalent single capacitor of capacitance

$$\frac{(4 \ \mu\text{F})(16 \ \mu\text{F})}{(4 \ \mu\text{F} + 16 \ \mu\text{F})} = 3 \ \mu\text{F}.$$

We then have three capacitors in parallel: a 3 μF, a 20 μF, and a 10 μF capacitor. Their equivalent capacitance is

$$3 \ \mu\text{F} + 20 \ \mu\text{F} + 10 \ \mu\text{F} = 33 \ \mu\text{F}.$$

18.33

a) Use the definition of capacitance:

$$C = \left| \frac{Q}{V} \right| \implies |Q| = C\,|V| = (10 \times 10^{-6}\ \text{F})(120\ \text{V}) = 1.2 \times 10^{-3}\ \text{C}.$$

b) Capacitors in series all the have the same magnitude charge on them. Hence, to store additional charge, connect additional capacitors in *parallel* with the given capacitor.

c) Each 10 μF capacitor in parallel has the same potential difference across it and, therefore, the same charge. To store a total of 1.0 C with each capacitor holding 1.2×10^{-3} C, requires

$$\frac{1.0\ \text{C}}{1.2 \times 10^{-3}\ \text{C}} = 8.3 \times 10^2$$

capacitors.

18.37 The equivalent capacitance of a parallel connection is the sum of the capacitances. Therefore, since you want an equivalent capacitance *smaller* than the one you have, a parallel connection is *not* the one to use. For a series connection of capacitors, the equivalent capacitance is smaller than the smallest in the collection, so a series connection is the one to use. Let C be the capacitance of an unknown capacitor to be placed in series with a single 15 μF capacitor. Then C has to satisfy the equation

$$10\ \mu\text{F} = \frac{(15\ \mu\text{F})C}{15\ \mu\text{F} + C} \implies C = 30\ \mu\text{F}.$$

Isn't that lucky! You can get the equivalent of one 30 μF capacitor by connecting two 15 μF capacitors in parallel! Here's the final arrangement of the three 15 μF capacitors.

It would really make sense, though, to go out and buy a few 10 μF capacitors. Each one costs quite a bit less than a 15 μF capacitor, and much less than three 15 μF capacitors.

18.41 Let C be the unknown capacitance and $V = 120$ V the potential difference across the capacitor. Then its potential energy is

$$\text{PE} = \frac{1}{2}CV^2 \implies C = \frac{2\text{PE}}{V^2} = \frac{2(1.0\ \text{J})}{(120\ \text{V})^2} = 1.4 \times 10^{-4}\ \text{F} = 1.4 \times 10^2\ \mu\text{F}.$$

18.45 The electrical potential energy is

$$\text{PE} = \frac{1}{2}C_{\text{eq}}V^2,$$

where V is the potential difference across the equivalent capacitance C_{eq}. Hence, to have the greatest potential energy, the equivalent capacitance must be the greatest.

The two capacitors have the same capacitance C. If we put them in series, then

$$\frac{1}{C_{\text{eq}}} = \frac{1}{C} + \frac{1}{C} = \frac{2}{C} \implies C_{\text{eq}} = \frac{C}{2}.$$

If we put them in parallel, then

$$C_{\text{eq}} = C + C = 2C.$$

Since $2C > \dfrac{C}{2}$, hook them in parallel. Here's how:

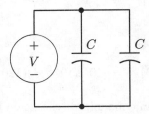

Then the potential energy stored is

$$PE = \frac{1}{2}(2C)V^2 = CV^2.$$

18.49   The potential energy stored is

$$PE = \frac{1}{2}CV^2.$$

The independent voltage source maintains the same potential difference across the plates of the capacitor as the dielectric slab is inserted. However, once inserted, the effect of the slab is to increase the capacitance from C to κC. Hence, the stored potential energy increases by the factor κ.

18.53   Each half of the capacitor has the same potential difference across it. Hence, the arrangement is equivalent to a parallel combination of two capacitors, each with half the area of the given capacitor. The two capacitors have capacitances

$$C_1 = \kappa_1 \frac{\left(\epsilon_0 \dfrac{A}{2}\right)}{d} = \frac{\kappa_1}{2}\left(\frac{\epsilon_0 A}{d}\right) \quad \text{and} \quad C_2 = \kappa_2 \frac{\left(\epsilon_0 \dfrac{A}{2}\right)}{d} = \frac{\kappa_2}{2}\left(\frac{\epsilon_0 A}{d}\right).$$

The equivalent capacitance of their parallel combination is

$$C_{eq} = C_1 + C_2 = \frac{\kappa_1}{2}\left(\frac{\epsilon_0 A}{d}\right) + \frac{\kappa_2}{2}\left(\frac{\epsilon_0 A}{d}\right) = \frac{\epsilon_0 A}{d}\left(\frac{\kappa_1 + \kappa_2}{2}\right).$$

18.57   Model the system as a parallel plate capacitor. The capacitance of a parallel plate capacitor with area A and plate separation d is

$$C = \frac{\epsilon_0 A}{d} = \frac{1}{4\pi\left(\dfrac{1}{4\pi\epsilon_0}\right)}\frac{A}{d} = \frac{1}{4\pi(9.00 \times 10^9 \text{ N·m}^2/\text{C}^2)}\frac{25 \times 10^6 \text{ m}^2}{2 \times 10^3 \text{ m}} = 1 \times 10^{-7} \text{ F} = 0.1 \ \mu\text{F}.$$

The dielectric strength of air is 3×10^6 V/m. This is the maximum magnitude E of the electric field between the two plates before the dielectric breaks down (lightning!). The potential difference V between the cloud and ground is related to the magnitude of the electric field strength $E = 3 \times 10^6$ V/m and the separation $d = 2$ km $= 2 \times 10^3$ m by

$$V = Ed = (3 \times 10^6 \text{ V/m})(2 \times 10^3) = 6 \times 10^9 \text{ V}.$$

The potential energy stored in the system is therefore

$$PE = \frac{1}{2}CV^2 = \frac{1}{2}(1 \times 10^{-7} \text{ F})(6 \times 10^9 \text{ V})^2 = 2 \times 10^{12} \text{ J}.$$

It is interesting that the capacitance of the system is only about 0.1 μF. Capacitors with this capacitance are very common in electronic equipment, and cost very little. On the other hand, capacitors with a breakdown voltage as high as 6×10^9 V (six *billion* volts), are definitely *not* commercially available!

Chapter 19

Electric Current, Resistance, and DC Circuit Analysis

19.1   The current is the product of the number of electrons per unit time and the magnitude of the charge on each, so

$$I = (10^6 \text{ electron/s})(1.602 \times 10^{-19} \text{ C/electron}) = 1.6 \times 10^{-13} \text{ C/s} \approx 2 \times 10^{-13} \text{ A} = 0.2 \text{ pA}.$$

19.5

a)   The current I, the number n of charge carriers per unit volume, the charge q per charge carrier, the drift speed $\langle v \rangle$, and the cross-sectional area A are related by

$$I = nq\langle v \rangle A \implies \langle v \rangle = \frac{I}{nqA}.$$

We are given that

$$I = 1.50 \text{ A}.$$

To compute the number n of charge carriers, note that: the number of moles per cubic meter of silver is

$$\frac{10.5 \times 10^3 \text{ kg/m}^3}{0.108 \text{ kg/mol}} = 9.72 \times 10^4 \text{ mol/m}^3.$$

Since each mole has Avogadro's number, 6.02×10^{23}, of silver atoms, and each silver atom has one electron charge carrier, the number of charge carriers per cubic meter is

$$n = (9.72 \times 10^4 \text{ mol/m}^3)(6.02 \times 10^{23} \text{ charge carriers/mol}) = 5.85 \times 10^{28} \text{ charge carriers/m}^3.$$

Since the charge carriers are electrons, the magnitude of the charge on each carrier is

$$q = |-e| /\text{charge carrier} = 1.602 \times 10^{-19} \text{ C/charge carrier}.$$

From Table 19.2 on page 843 of the text, the cross-sectional area of 12 gauge wire is

$$A = 3.31 \times 10^{-6} \text{ m}^2.$$

Putting it all together, the drift speed is

$$\langle v \rangle = \frac{1.50 \text{ A}}{(5.85 \times 10^{28} \text{ charge carriers/m}^3)(1.602 \times 10^{-19} \text{ C/charge carrier})(3.31 \times 10^{-6} \text{ m}^2)}$$
$$= 4.84 \times 10^{-5} \text{ m/s}.$$

167

b)   The time a charge carrier takes to travel a distance d of exactly one meter is

$$t = \frac{d}{|\langle v \rangle|} = \frac{1\text{ m}}{4.84 \times 10^{-5}\text{ m/s}} = \left(\frac{1\text{ m}}{4.84 \times 10^{-5}\text{ m/s}} \right) \left(\frac{\text{h}}{3600\text{ s}} \right) = 5.74\text{ h}.$$

19.9   The resistance of the resistor is

$$R = \frac{\rho \ell}{A},$$

where ρ is the resistivity of gold, ℓ is the length of the resistor, and A is its cross-sectional area. From Table 19.1 on page 843 of the text, $\rho = 2.44 \times 10^{-8}\ \Omega \cdot \text{m}$. From the problem statement, $\ell = 400 \times 10^{-6}$ m and $A = (1.0 \times 10^{-6}\text{ m})(10.0 \times 10^{-6}\text{ m}) = 10 \times 10^{-12}\text{ m}^2$. Hence

$$R = \frac{(2.44 \times 10^{-8}\ \Omega \cdot \text{m})(400 \times 10^{-6}\text{ m})}{10 \times 10^{-12}\text{ m}^2} = 0.98\ \Omega.$$

19.13   View the pipeline as a big long resistor. Its resistance is

$$R = \frac{\rho \ell}{A},$$

where ρ is the resistivity of steel, ℓ is the length of the pipeline, and A is its cross-sectional area (but just the steel part of the cross section — don't include the hollow part that ordinarily carries the oil).

We're given the resistivity of steel as $\rho = 1.80 \times 10^{-7}\ \Omega \cdot \text{m}$, and we're given $\ell = 1.27 \times 10^6$ m.

It remains to find A. Consider a cross section of the pipe. The steel portion of the cross section looks like a washer. Cut the washer at one point, and straighten it out. You'll get a strip of steel whose length is the circumference of the pipe, $\pi(1.20\text{ m})$, and whose width is the thickness of the pipe's wall, 0.010 m. Hence, $A = \pi(1.20\text{ m})(0.010\text{ m}) = 0.038\text{ m}^2$.

Thus,

$$R = \frac{(1.80 \times 10^{-7}\ \Omega \cdot \text{m})(1.27 \times 10^6\text{ m})}{0.038\text{ m}^2} = 6.0\ \Omega.$$

19.17   From Equation 19.3 on page 840 of the text, the current in the big wire is

$$I = nq \langle v \rangle A \implies \langle v \rangle = \frac{I}{nqA},$$

where n is the number of charge carriers per unit volume, q is the amount of charge per carrier, $\langle v \rangle$ is the drift speed, and A is the cross-sectional area. Let d be its diameter. Then $A = \pi \left(\dfrac{d}{2} \right)^2$, so

$$\langle v \rangle = \frac{I}{nq\pi \left(\dfrac{d}{2} \right)^2} = \frac{4I}{nq\pi d^2}.$$

For the small wire, I, n, and q are the same, but the diameter is $\dfrac{d}{2}$, so the cross-sectional area is $A' = \pi \left(\dfrac{d}{4} \right)^2$. Therefore its drift speed will be

$$\langle v' \rangle = \frac{I}{nq\pi \left(\dfrac{d}{4} \right)^2} = \frac{16I}{nq\pi d^2}.$$

The drift speed in the small wire is greater than that in the big wire by a factor of 4.

19.21 The resistance is

(1)
$$R = \frac{\rho\ell}{A},$$

where ρ (without any subscript) is the resistivity. The mass is

$$m = \rho_{\text{mass}}V,$$

where ρ_{mass} is the mass density (i.e., the mass per unit volume), and V is the volume. But the volume is the length times the cross-sectional area, $V = \ell A$, so

$$m = \rho_{\text{mass}}A\ell \implies A = \frac{m}{\ell\rho_{\text{mass}}}.$$

Substitute this expression for A into equation (1) above,

$$R = \frac{\rho\ell}{\left(\dfrac{m}{\ell\rho_{\text{mass}}}\right)} = \rho\ell^2\rho_{\text{mass}}\left(\frac{1}{m}\right).$$

So the resistance is inversely proportional to the mass, and the proportionality constant is $\rho\ell^2\rho_{\text{mass}}$.

19.25

a)   The two $1.0\text{ k}\Omega$ resistors on the far right of the network are in series, so we may replace them by a single $1.0\text{ k}\Omega + 1.0\text{ k}\Omega = 2.0\text{ k}\Omega$ resistor. Here's the new network.

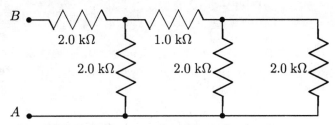

The two $2.0\text{ k}\Omega$ resistors on the far right are in parallel, so they may be replaced by a single resistor of resistance

$$\frac{(2.0\text{ k}\Omega)(2.0\text{ k}\Omega)}{2.0\text{ k}\Omega + 2.0\text{ k}\Omega} = 1.0\text{ k}\Omega.$$

The new network looks like

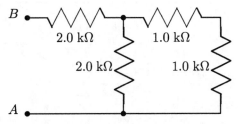

The two $1.0\text{ k}\Omega$ resistors on the far right are in series, so they may be replaced by a single resistor with resistance $1.0\text{ k}\Omega + 1.0\text{ k}\Omega = 2.0\text{ k}\Omega$. After doing this, the result looks like

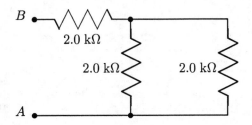

The two 2.0 kΩ resistors on the far right are in parallel. They may be replaced by a single resistor with resistance

$$\frac{(2.0 \text{ k}\Omega)(2.0 \text{ k}\Omega)}{2.0 \text{ k}\Omega + 2.0 \text{ k}\Omega} = 1.0 \text{ k}\Omega.$$

After doing this, the new network is

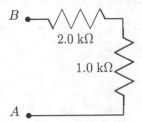

Finally, the two remaining resistors are in series. They may be replaced by a single 2.0 kΩ + 1.0 kΩ = 3.0 kΩ resistor. The final network is

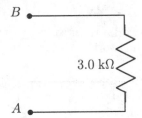

b)   The two 2.0 Ω resistors on the far right of the network are in series. Replace them with a single 2.0 Ω + 2.0 Ω = 4.0 Ω resistor. After doing this, we have

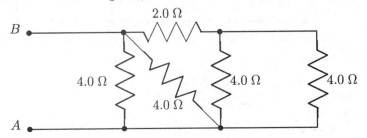

The two 4.0 Ω resistors on the far right are in parallel. Replace them by a single resistor with resistance

$$\frac{(4.0 \text{ }\Omega)(4.0 \text{ }\Omega)}{4.0 \text{ }\Omega + 4.0 \text{ }\Omega} = 2.0 \text{ }\Omega.$$

After doing this, the new network is

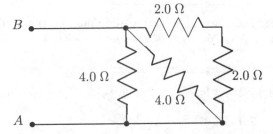

The two 2.0 Ω resistors on the far right are in series. Replace them by a single resistor with resistance 2.0 Ω + 2.0 Ω = 4.0 Ω After doing this, the new network is

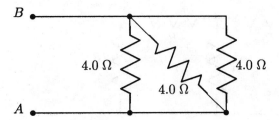

The two 4.0 Ω resistors on the far right are in parallel. Replace them by a single resistor with resistance

$$\frac{(4.0\ \Omega)(4.0\ \Omega)}{4.0\ \Omega + 4.0\ \Omega} = 2.0\ \Omega.$$

After doing this, the new network looks like

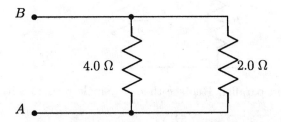

The two remaining resistors are in parallel. Replace them by a single resistor with resistance

$$\frac{(4.0\ \Omega)(2.0\ \Omega)}{4.0\ \Omega + 2.0\ \Omega} = 1.3\ \Omega.$$

After doing this, the final network is

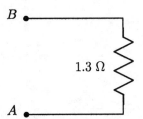

c)   The two resistors on the bottom right of the network are in series. Replace them by a $R + R = 2R$ resistor. The result is

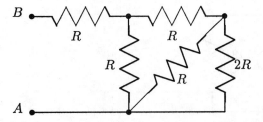

The two resistors on the bottom right are in parallel. Replace them by a resistor with resistance equivalent

$$R = \frac{(R)(2R)}{R + 2R} = \frac{2R}{3}.$$

The result is

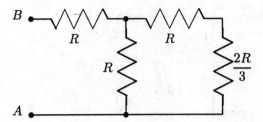

The two resistors on the far right are in series. Replace them by a resistor with resistance $R + \dfrac{2R}{3} = \dfrac{5R}{3}$. After doing this, we have

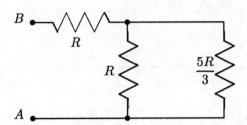

The two resistors on the right are in parallel. Replace them by single resistor with resistance

$$\frac{(R)\left(\dfrac{5R}{3}\right)}{R + \dfrac{5R}{3}} = \frac{5R}{8}.$$

The result is

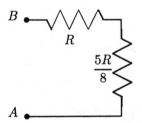

These two resistors are in series, so they may be replaced by a single resistor with resistance $R + \dfrac{5R}{8} = \dfrac{13R}{8}$, leaving us with

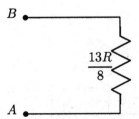

d)   The two $4.0\,\Omega$ resistors are in parallel. Replace them by the single resistor with resistance

$$\frac{(4.0\,\Omega)(4.0\,\Omega)}{4.0\,\Omega + 4.0\,\Omega} = 2.0\,\Omega.$$

The result is

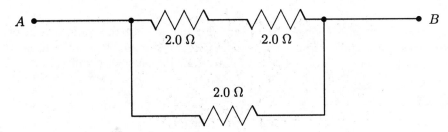

The upper two resistors are in series. Replace them by a single $2.0\ \Omega + 2.0\ \Omega = 4.0\ \Omega$ resistor. The result is

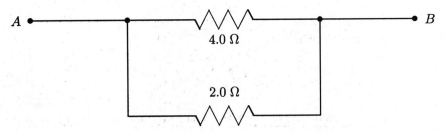

The remaining two resistors are in parallel. Replace them by a single resistor with resistance

$$\frac{(4.0\ \Omega)(2.0\ \Omega)}{4.0\ \Omega + 2.0\ \Omega} = 1.3\ \Omega.$$

The result is

A •————————$\sim\!\!\!\!\sim\!\!\!\!\sim$————————• B
$1.3\ \Omega$

19.29 Given a potential difference V and current I, the power absorbed is

(1) $$P = IV.$$

For a given resistance R and current I, we can use Ohm's law to find the potential difference across the resistance,

$$V = IR \implies I = \frac{V}{R}.$$

Substitute this expression for I into equation (1).

$$P = \frac{V^2}{R} \iff R = \frac{V^2}{P}.$$

Therefore, for a fixed potential difference, the power and the resistance are *inversely* proportional to each other. Since both bulbs have the same potential difference $V = 120$ V across them, the bulb that absorbs the least power is the one with the greatest resistance. The 100 W bulb has greater resistance than the 200 W bulb, since

$$R_{100\ \text{W bulb}} = \frac{(120\ \text{V})^2}{100\ \text{W}} = 144\ \Omega \quad > \quad R_{200\ \text{W bulb}} = \frac{(120\ \text{V})^2}{200\ \text{W}} = 72\ \Omega.$$

19.33

a)   View a piece of wire one meter long as a resistor. First find what the resistance R of the piece of wire has to be in order to absorb 1.0 W of power when carrying a current of 15 A. Then use the resistivity ρ of copper from Table 19.1 on page 843 of the text to find what diameter wire will result in this resistance.

The power absorbed by the piece of wire is

(1) $$P = IV,$$

where I is the current through it and V is the potential difference between the two ends. According to Ohm's law, the potential difference is

$$V = IR.$$

Substitute this expression for V into equation (1) and solve for R:

$$P = I(IR) = I^2 R \implies R = \frac{P}{I^2} = \frac{1.0 \text{ W}}{(15 \text{ A})^2} = 4.4 \times 10^{-3} \ \Omega.$$

The resistance of the piece of wire is related to its resistivity ρ, length ℓ, and cross-sectional area A by

(2) $$R = \frac{\rho \ell}{A}.$$

The cross-sectional area $A = \pi \left(\dfrac{d}{2} \right)^2$. Substitute this expression for A into equation (2) and solve for d.

$$R = \frac{\rho \ell}{\pi \left(\dfrac{d}{2} \right)^2} = \frac{4\rho \ell}{\pi d^2} \implies d = \sqrt{\frac{4\rho \ell}{\pi R}}.$$

From Table 19.1 (on page 843 of the text), the resistivity of copper is $\rho = 1.77 \times 10^{-8} \ \Omega{\cdot}\text{m}$. Use this together with $\ell = 1.0$ m and the value $R = 4.4 \times 10^{-3} \ \Omega$ found above to compute

$$d = \sqrt{\frac{4(1.77 \times 10^{-8} \ \Omega{\cdot}\text{m})(1.0 \text{ m})}{\pi(4.4 \times 10^{-3} \ \Omega)}} = 2.3 \times 10^{-3} \text{ m}.$$

b)   The diameter of the wire must be no less than the value $d = 2.3 \times 10^{-3}$ m calculated in part a). Hence, according to Table 19.2 on page 843 of the text, the maximum gauge number that can be used is 10 gauge wire, which has a diameter of 2.59×10^{-3} m. (It's a peculiarity of the "gauge" measurement system that the smaller the gauge, the bigger the wire.)

19.37   In order to see what happens when we put the light bulbs in series, we'll need their resistances, so we'll compute them first. The power absorbed by each bulb is

(1) $$P = IV,$$

where I is the current through the bulb and V is the potential difference across it. According to Ohm's law, if the resistance is R, then

$$V = IR \implies I = \frac{V}{R}.$$

Substitute this expression for I into equation (1) and solve for R:

$$P = \frac{V^2}{R} \implies R = \frac{V^2}{P}.$$

The power rating for light bulbs is given under the assumption that they will be hooked up normally — in this case, with 120 V across them. Thus we may use $V = 120$ V in the last equation along with the bulbs' power ratings for P, to determine their resistances:

$$R_{100\,W} = \frac{(120\text{ V})^2}{100\text{ W}} = 144\ \Omega \qquad \text{and} \qquad R_{60\,W} = \frac{(120\text{ V})^2}{60\text{ W}} = 2.4 \times 10^2\ \Omega\,.$$

Now let's hook them up in series with a 120 V source.

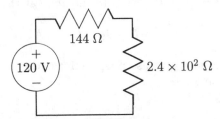

The 100 W bulb is at the top, the 60 W bulb is on the right. In order to find the power absorbed by each, we need to find the current through each. The current through each one will be the same — its the current in the circuit. The two bulbs are in series, so their resistances add. They are equivalent to a single $R = 144\ \Omega + 2.4 \times 10^2\ \Omega = 3.8 \times 10^2\ \Omega$ resistor. Thus we may use Ohm's law to compute the current I in the circuit:

$$V = IR \implies I = \frac{V}{R} = \frac{120\text{ V}}{3.8 \times 10^2\ \Omega} = 0.32\text{ A}\,.$$

The power absorbed by a resistor is $P = IV$, where I is the current through it, and V is the potential difference across it. From Ohm's law, $V = IR$, so we may rewrite the power as $P = I^2 R$. Applying this to each light bulb we have

$$P_{100\,W} = (0.32\text{ A})^2 (144\ \Omega) = 15\text{ W} \qquad \text{and} \qquad P_{60\,W} = (0.32\text{ A})^2 (2.4 \times 10^2\ \Omega) = 25\text{ W}\,.$$

The 60 W bulb burns more brightly since it absorbs more power!

At first it may seem strange that the "less powerful" bulb absorbs the most power (converts more electrical power to heat transfer and light). But this is because we have hooked them up in series, rather than in the usual parallel arrangement.

When in series, the current through both bulbs will be the same, and the potential difference across the higher resistance bulb will be the greatest so it will absorb more power. In parallel, the potential difference across both bulbs is the same, and the current through the higher resistance bulb will be less than the low resistance bulb, so the former will absorb less power.

19.41

a) and

b) We'll use the current directions indicated in the sketch below. The resistor polarities are chosen consistent with the current direction so current enters each resistor at its + polarity terminal, and leaves at its − polarity terminal.

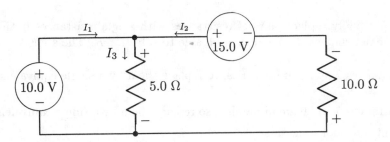

c) Apply the KVL to each little (elementary) loop, going in a clockwise direction and adding up the potential differences, +(potential difference at + terminal) − (potential difference at − terminal). In each loop, we've begun with the voltage source:

(Left loop) $$-10.0 \text{ V} + (5.00 \, \Omega) I_3 = 0 \text{ V}$$
(Right loop) $$15.0 \text{ V} - (10.0 \, \Omega) I_2 - (5.00 \, \Omega) I_3 = 0 \text{ V}$$

This gives us two equations. Now apply the KCL to the node at the top of the diagram.

(KCL) $$-I_1 - I_2 + I_3 = 0 \text{ A}.$$

These three (simultaneous) equations give us enough information to find the three currents I_1, I_2, and I_3. First solve the Left loop equation for I_3:

$$I_3 = \frac{10.0 \text{ V}}{5.0 \, \Omega} = 2.0 \text{ A}.$$

Now substitute this value for I_3 into the Right loop equation and solve for I_2:

$$15.0 \text{ V} - (10.0 \, \Omega) I_2 - (5.00 \, \Omega)(2.0 \text{ A}) = 0 \text{ V} \implies I_2 = 0.50 \text{ A}.$$

Finally, substitute these values for I_3 and I_2 into (KCL) and solve for I_1:

$$-I_1 - 0.50 \text{ A} + 2.0 \text{ A} = 0 \text{ A} \implies I_1 = 1.5 \text{ A}.$$

d) The power absorbed by (or transferred to) each circuit element is the product of the current into the positive polarity terminal times the potential difference across the element. The power absorbed by the 10.0 V battery is

$$P_{10.0 \text{ V}} = (-1.5 \text{ A})(10.0 \text{ V}) = -15 \text{ W}.$$

The power absorbed by the 15.0 V battery is

$$P_{15.0 \text{ V}} = (-0.50 \text{ A})(15.0 \text{ V}) = -7.5 \text{ W}.$$

The power absorbed by the 5.0 Ω resistor is

$$P_{5.0 \, \Omega} = (2.0 \text{ A})[(2.0 \text{ A})(5.0 \, \Omega)] = 20 \text{ W}.$$

The power absorbed by the 10.0 Ω resistor is

$$P_{10.0 \, \Omega)} = (0.50 \text{ A}[(0.50 \text{ A})(10.0 \, \Omega)] = 2.5 \text{ W}.$$

e) Note that the sum of all the power absorbed is indeed 0 W.

19.45

a) We'll gradually replace pairs of resistances with single resistances until we're left with one resistance whose value depends on R. We'll then be able to solve for R. The steps we follow are:

1. Resistances 2 and 3 are in series, so replace them by a single equivalent resistance $R + R = 2R$.

2. Resistances 5 and 6 are in parallel, so replace them by a single equivalent resistance $\frac{RR}{R+R} = \frac{R}{2}$. The circuit now looks like this:

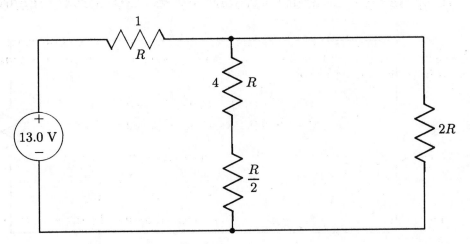

3. Now Resistance 4 is in series with the new $\frac{R}{2}$ resistance, so replace the two of them with an equivalent $R + \frac{R}{2} = \frac{3R}{2}$ resistance.

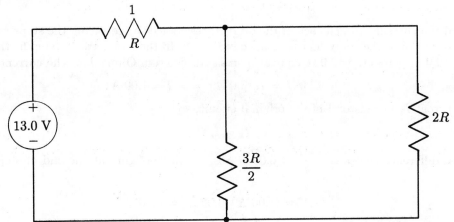

4. Now the $2R$ and $\frac{3R}{2}$ resistances are in parallel, so replace them by a single $\dfrac{2R\frac{3R}{2}}{R + \frac{3R}{2}} = \dfrac{6R}{7}$ resistance.

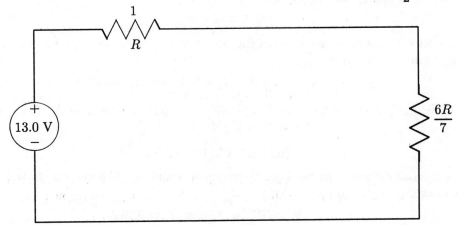

5. The only resistances left are the newly created $\frac{6R}{7}$ resistance and the old resistance R. Since they are in series, replace them with an equivalent $\frac{6R}{7} + R = \frac{13R}{7}$ resistance.

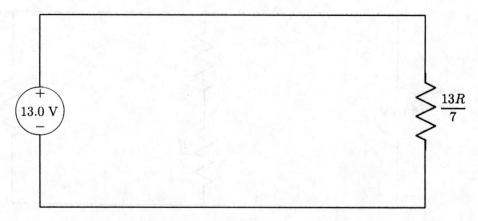

The equivalent resistance is given to be $13.0\,\Omega$, so

$$13.0\,\Omega = \frac{13R}{7} \implies R = 7.00\,\Omega\,.$$

b)   and

c)   We'll find the current through each of the original resistances by retracing the steps we took in part a).
From Zeus, we know the equivalent resistance is $13.0\,\Omega$. In the last network (step 5) there is a potential
difference of $13.0\,V$ across the $13.0\,\Omega$ equivalent resistance. From Ohm's law, the current satisfies

$$13.0\,V = I(13.0\,\Omega) \implies I = 1.00\,A\,.$$

This is the current I_1 in resistor 1 of the original circuit, so

$$I_1 = 1.00\,A\,.$$

The potential difference across the $\dfrac{6R}{7} = 6.00\,\Omega$ resistor in the figure at the end of step 4 is found from
Ohm's law:

$$V = (1.00\,A)(6.00\,\Omega) = 6.00\,V\,.$$

Therefore the potential difference across the $2R = 14.00\,\Omega$ resistance in the figure at the end of step 3 is
$6.00\,V$. So from Ohm's law, the current I through it satisfies

$$6.00\,V = I(14.00\,\Omega) \implies I = 0.429\,A\,.$$

This is the current I_2 and I_3 through Resistances 2 and 3, so

$$I_2 = I_3 = 0.429\,A\,.$$

Now apply the KCL to the node at the top of the circuit at the end of step 2, let I_4 be the current through
Resistance 4. Then

$$-1.00\,A + 0.429\,A + I_4 = 0\,A \implies I_4 = 0.57\,A\,.$$

From Ohm's law, the potential difference across the $\dfrac{R}{2} = 3.50\,\Omega$ resistance in the figure at the end of step
2 is

$$(0.57\,A)(3.50\,\Omega) = 2.0\,V\,.$$

This is the potential difference across both Resistances 5 and 6. The current I_5 through Resistance 5
therefore satisfies

$$2.0\,V = I_5(7.00\,\Omega) \implies I_5 = 0.29\,A\,.$$

Similarly,

$$I_6 = 0.29\,A\,.$$

Hence the smallest current, $0.29\,A$, is through Resistances 5 and 6.

d)   To increase the current provided by the battery, decrease the effective equivalent resistance of the entire circuit. A set of resistances in parallel has an equivalent resistance smaller than the smallest in the parallel connection. Hence, to decrease the equivalent resistance in figure (4) above with a single resistance R, place R in parallel with the battery and the rest of the circuit, as shown below:

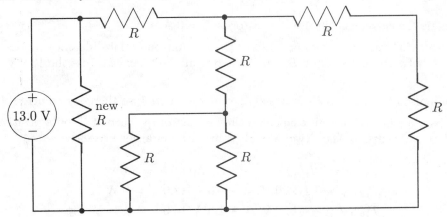

19.49

a)   Choose the current directions as shown below. The resistor's polarities are marked so that current enters each resistor at the + polarity terminal.

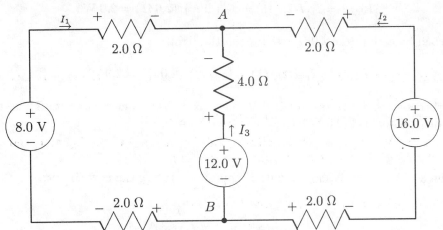

There are three quantities to be determined, I_1, I_2, and I_3, so we need three (independent) equations. Two of these can be obtained by applying KVL to the left and right loops in the above circuit. A third can be obtained by applying KCL to node A. Here are the three equations (For the KVL equations we traverse each loop clockwise, beginning at node A.):

(KVL left loop)	$-(4.0\,\Omega)I_3 + 12.0\text{ V} + (2.0\,\Omega)I_1 - 8.0\text{ V} + (2.0\,\Omega)I_1 = 0\text{ V}$
(KVL right loop)	$-(2.0\,\Omega)I_2 + 16.0\text{ V} - (2.0\,\Omega)I_2 - 12.0\text{ V} + (4.0\,\Omega)I_3 = 0\text{ V}$
(KCL node A)	$-I_1 - I_2 - I_3 = 0\text{ A}$

Rewrite these equations more neatly, by combining the coefficients of each of the variables, putting the variables in the same order on the left side, and combining the constant terms on the right.

(KVL left loop)	$(4.0\,\Omega)I_1$		$-(4.0)I_3$	$= -4.0\text{ V}$
(KVL right loop)		$-(4.0\,\Omega)I_2$	$+(4.0\,\Omega)I_3$	$= -4.0\text{ V}$
(KCL node A)	$-I_1$	$-I_2$	$-I_3$	$= 0\text{ A}$

These are three simultaneous linear equations in the three unknowns I_1, I_2, and I_3. Solve them in any one of a number of ways. For example, solve the first equation for I_1 in terms of I_3 to obtain $I_1 = I_3 - 1.0\text{ A}$.

Solve the second for I_2 in terms of I_3, to obtain $I_2 = I_3 + 1.0$ A. Substitute these expressions for I_1 and I_2 into the third equation, and find that $-I_3 - I_3 - I_3 = 0$ A, so $I_3 = 0$ A. Therefore,

$$I_1 = -1.0 \text{ A}, \qquad I_2 = 1.0 \text{ A}, \qquad \text{and} \qquad I_3 = 0 \text{ A}.$$

Hence, in particular, the current in the 4.0 Ω resistor is $I_3 = 0$ A.

b) Since the current in the 4.0 Ω resistor is zero, there is no potential difference across it. Hence the potential difference between points A and B is just the potential difference across the 12.0 V battery, which is 12.0 V.

c) The two terminals of the 4.0 Ω resistor are at the same potential, so there is no current in this resistor.

d) The power absorbed by each circuit element is the product of the current into its positive terminal and the potential difference across it. The power absorbed by each voltage source is

$$P_{8.0 \text{ V}} = -I_1 8.0 \text{ V} = -(-1.0 \text{ A})8.0 \text{ V} = 8.0 \text{ W}.$$
$$P_{12.0 \text{ V}} = -I_3 12.0 \text{ V} = -(-0 \text{ A})12.0 \text{ V} = 0 \text{ W}.$$
$$P_{16.0 \text{ V}} = -I_2 16.0 \text{ V} = -(1.0 \text{ A})16.0 \text{ V} = -16 \text{ W}.$$

The total power absorbed by all the voltage sources is -8.0 W.

e) The power absorbed by each of the 2.0 Ω resistors on the left is

$$P_{2.0\,\Omega} = I_1(I_1 2.0\,\Omega) = (-1.0 \text{ A})^2(2.0\,\Omega) = 2.0 \text{ W}.$$

The power absorbed by each of the 2.0 Ω resistors on the right is

$$P_{2.0\,\Omega} = I_1(I_1 2.0\,\Omega) = (1.0 \text{ A})^2(2.0\,\Omega) = 2.0 \text{ W}.$$

The power absorbed by the 4.0 Ω resistor is 0 W because it has no current in it. Thus, the total power absorbed by all the resistors is $2(2.0 \text{ W}) + 2(2.0 \text{ W}) + 0 \text{ W} = 8.0 \text{ W}$.

19.53

a) Note that the 3.0 Ω resistor is shorted out. Eliminate it. Here's the new circuit.

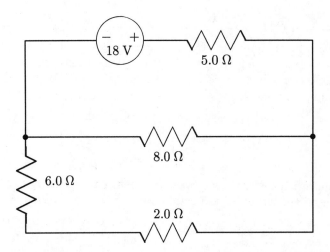

The 2.0 Ω and 6.0 Ω resistors are in series. Replace them with their equivalent 2.0 Ω + 6.0 Ω = 8.0 Ω resistor.

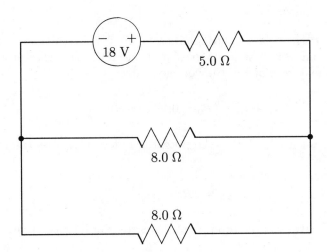

The two 8.0 Ω resistors are in parallel. Replace them by an equivalent $\dfrac{(8.0\ \Omega)(8.0\ \Omega)}{8.0\ \Omega + 8.0\ \Omega} = 4.0\ \Omega$ resistor.

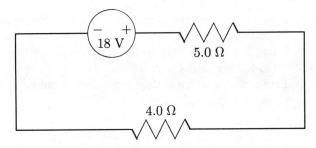

The 5.0 Ω and 4.0 Ω resistors are in series. Replace them by a single $5.0\ \Omega + 4.0\ \Omega = 9.0\ \Omega$ resistor.

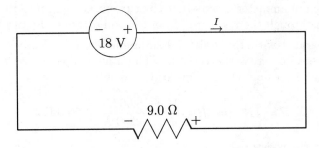

b)   The current direction and the resulting resistor polarity are indicated in the sketch above. Apply the KVL clockwise around the loop:

$$-18\text{ V} + I(9.0\ \Omega) = 0\text{ V} \implies I = 2.0\text{ A}.$$

c)   Notice that in the second diagram in part a), the two 8.0 Ω resistors are in parallel. Because the two resistances are equal, the current divides equally between them. The current in the 8.0 Ω resistor is therefore 1.0 A. The power absorbed by this resistor is

$$P_{8.0\ \Omega} = IV = I[I(8.0\ \Omega)] = (1.0\text{ A})(1.0\text{ A})(8.0\ \Omega) = 8.0\text{ W}.$$

d)   The power absorbed by the independent voltage source is the product of the current into its positive terminal and the potential difference, so

$$P_{18\text{ V}} = (-2.0\text{ A})(18\text{ V}) = -36\text{ W}.$$

19.57

a) If R is infinite, then no current passes through its section of the circuit, and we are left with the simple circuit consisting of V_0, r, and R in a series loop. Apply the KVL clockwise around this loop, taking the current I to emerge from the $+$ terminal of the voltage source:

$$-V_0 + Ir + IR_0 = 0 \text{ V} \implies I = \frac{V_0}{r + R_0}.$$

The potential difference between A and B is the potential difference across the resistor R_0, which we can find with Ohm's law:

$$V_{AB} = IR_0 = \frac{V_0}{r + R_0} R_0.$$

The power absorbed by the battery is the product of the current into its positive terminal and it's potential difference:

$$P = (-I)V_0 = \left(-\frac{V_0}{r + R_0}\right)V_0 = -\frac{V_0^2}{r + R_0}.$$

b) If the resistance R is zero, then we can replace it with a resistance-free wire. Then R_0 is shorted out and can be eliminated from the circuit. We are left with the simple circuit consisting only of V_0 and r in a series loop. (Essentially, we are left with a *real* battery whose terminals have been connected by a resistance free wire. Note that this is *not* a good thing to do in practice!)

   Now the potential difference between points A and B is zero. Apply the KVL clockwise around the loop, taking the current to be in the same direction as before:

$$-V_0 + Ir = 0 \text{ V} \implies I = \frac{V_0}{r}$$

19.61

a) The positive reading of the ammeter means that the current is going into its positive terminal. Hence, the direction of the current through the resistor is from top to bottom in Figure P.61.

b) The voltmeter is in parallel with the resistor (the ammeter has effectively zero resistance), and so indicates the potential difference across the resistor. The ammeter is in series with the resistor and so indicates the current through it. Apply Ohm's law to the resistor:

$$V = IR \implies R = \frac{V}{I} = \frac{5.00 \text{ V}}{0.250 \text{ A}} = 20.0 \ \Omega.$$

c) The power absorbed by the resistor is

$$P = IV = (0.250 \text{ A})(5.00 \text{ V}) = 1.25 \text{ W}.$$

d) Because the voltmeter is in parallel with the independent voltage source, the two have the same potential difference across them. Hence, the emf of the source is 5.00 V.

19.65

a) With the switch open, there is no current through it, or anywhere to the right of it. The voltmeter V_1 has a huge effective resistance so there is essentially no current through r. In other words there is no current anywhere in the network. Since there is no current through r, there is no potential difference across it, so the positive terminals of the voltage source V_0 and the voltmeter V_1 are at the same potential. Their negative terminals are also at the same potential, since they are directly connected. Thus, the reading of the voltmeter V_0 is the emf of the independent voltage source, so $V_0 = 10.0 \text{ V}$.

   The two ammeters both indicate 0 A, because there is no current anywhere in the network.

b) With the switch closed, replace it, for purposes of analysis, by a piece of resistance free wire.

Begin with R_3. The potential difference across it is 6.0 V (the reading of the voltmeter V_2 in parallel with it) and the current through it is 0.75 A (the reading of the ammeter A_2 in series with it). Apply Ohm's law to find R_3.

$$V = IR_3 \implies R_3 = \frac{V}{I} = \frac{6.0 \text{ V}}{0.75 \text{ A}} = 8.0 \text{ }\Omega.$$

The potential difference across R_2 is also 6.0 V (the reading of the voltmeter V_2 in parallel with it) and the current through it is 0.50 A (the reading of ammeter A_1 in series with it). Apply Ohm's law to find R_2.

$$V = IR_2 \implies R_2 = \frac{V}{I} = \frac{6.0 \text{ V}}{0.50 \text{ A}} = 12 \text{ }\Omega.$$

The potential difference across R_1 is the difference in the readings of the two voltmeters, $8.0 \text{ V} - 6.0 \text{ V} = 2.0 \text{ V}$. From the KCL, the current through R_1 (and also r) is the sum of the two ammeter readings, $0.75 \text{ A} + 0.50 \text{ A} = 1.25 \text{ A}$. Apply Ohm's law to find R_1.

$$V = IR_1 \implies R_1 = \frac{V}{I} = \frac{2.0 \text{ V}}{1.25 \text{ A}} = 1.6 \text{ }\Omega.$$

From part a), V_0 is a 10.0 V source. Because voltmeter V_1 reads 8.0 V when the switch is closed, the potential difference across r is $10.0 \text{ V} - 8.0 \text{ V} = 2.0 \text{ V}$. The current through r is the same as the current through R_1 which we found to be 1.25 A. Apply Ohm's law to find r:

$$V = Ir \implies r = \frac{V}{I} = \frac{2.0 \text{ V}}{1.25 \text{ A}} = 1.6 \text{ }\Omega.$$

19.69 Equation 19.41 on page 875 of the text describes how the charge on a discharging capacitor varies with time:

$$q(t) = Q_0 e^{-t/\tau},$$

where $\tau = RC$, is the time constant. When the charge is half the initial value, this equation becomes

$$\frac{Q_0}{2} = Q_0 e^{-t/\tau} \implies \frac{1}{2} = e^{-t/\tau} \implies \ln \frac{1}{2} = -\frac{t}{\tau} \implies t = -\left(\ln \frac{1}{2}\right)\tau = (\ln 2)\,\tau \approx 0.693\tau.$$

When the capacitor is charging (see page 874 of the text, just below Equation 19.39),

$$q(t) = Q_0 \left(1 - e^{-t/\tau}\right),$$

so, when $t = \tau \ln 2$,

$$q = Q_0 \left(1 - e^{-\ln 2}\right) = Q_0 \left(1 - \frac{1}{2}\right) = \frac{Q_0}{2}.$$

So when discharging, the capacitor takes the same amount of time to lose half of its full charge as it does to gain it when charging.

19.73 Each ideal capacitor acts as an open switch for dc currents. Hence, the circuit effectively looks like resistors R_1 and R_2 in a series loop with a voltage source V_0. Such a circuit is called a *voltage divider*. In fact, once the capacitors are removed we have precisely the same circuit as in Figure P.47 for problem 19.47.

First find the current I through the resistors. Since R_1 and R_2 are in series, their equivalent resistance is $R = R_1 + R_2$. The potential difference across them is V_0. Thus, by Ohm's law,

$$V_0 = I(R_1 + R_2) \implies I = \frac{V_0}{R_1 + R_2}.$$

Since R_1 is connected in parallel with C_1, the potential difference across the capacitor is the same as that across the resistor. From Ohm's law this is

$$V_1 = IR_1 = \left(\frac{R_1}{R_1 + R_2} \right) V_0.$$

A similar application of Ohm's law gives the potential difference across C_2 as

$$V_2 = IR_2 = \left(\frac{R_2}{R_1 + R_2} \right) V_0.$$

Chapter 20

Magnetic Forces and the Magnetic Field

20.1   The magnetic force on a charge q is

$$\vec{F} = q\vec{v} \times \vec{B}.$$

The sign of the charge is important. The force is perpendicular to both $\vec{v}$ and $\vec{B}$. For the positive charge, the direction of $\vec{F}$ is the same as the direction of $\vec{v} \times \vec{B}$; for the negative charge, the direction of $\vec{F}$ is *opposite* to the direction of $\vec{v} \times \vec{B}$. In either case, begin by using the vector product right-hand rule to determine the direction of the vector $\vec{v} \times \vec{B}$. The initial direction of the force on each charge as it enters the magnetic field is shown in the sketch below.

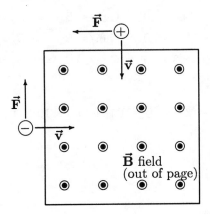

20.5   The magnetic force on a charge q is

$$\vec{F} = q\vec{v} \times \vec{B}.$$

The sign of the charge is important. The force is perpendicular to both $\vec{v}$ and $\vec{B}$. For the positive charge, the direction of $\vec{F}$ is the same as the direction of $\vec{v} \times \vec{B}$. For the negative charge, the direction of $\vec{F}$ is *opposite* to the direction of $\vec{v} \times \vec{B}$. In either case, begin by using the vector product right-hand rule to determine the direction of the vector $\vec{v} \times \vec{B}$. The direction of the force on the positive charge is out of the page. The direction of the force on the negative charge is into the page.

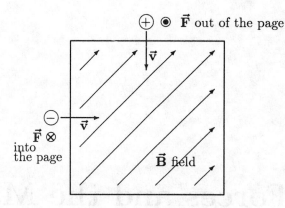

20.9   Since the magnetic field lacks an x-component, write the field as

$$\vec{\mathbf{B}} = B_y\hat{\mathbf{j}} + B_z\hat{\mathbf{k}}.$$

The magnetic force on the charge is

$$\vec{\mathbf{F}} = q\vec{\mathbf{v}} \times \vec{\mathbf{B}},$$

so

$$(3.84 \times 10^{-19}\ \text{N})\hat{\mathbf{i}} + (3.20 \times 10^{-19}\ \text{N})\hat{\mathbf{j}} + (2.40 \times 10^{-19}\ \text{N})\hat{\mathbf{k}}$$

$$= (-1.602 \times 10^{-19}\ \text{C})\left((5.0 \times 10^3\ \text{m/s})\hat{\mathbf{i}} - (6.0 \times 10^3\ \text{m/s})\hat{\mathbf{j}}\right) \times (B_y\hat{\mathbf{j}} + B_z\hat{\mathbf{k}})$$

$$= (-8.0 \times 10^{-16}\ \text{C·m/s})B_y\hat{\mathbf{k}} + (8.0 \times 10^{-16}\ \text{C·m/s})B_z\hat{\mathbf{j}} + (9.6 \times 10^{-16}\ \text{C·m/s})B_z\hat{\mathbf{i}}$$

$$= (9.6 \times 10^{-16}\ \text{C·m/s})B_z\hat{\mathbf{i}} + (8.0 \times 10^{-16}\ \text{C·m/s})B_z\hat{\mathbf{j}} - (8.0 \times 10^{-16}\ \text{C·m/s})B_y\hat{\mathbf{k}}.$$

Two vectors are equal to each other if and only if their respective components are equal to each other. Hence, equate corresponding components from the two sides of the above equation and solve for B_y and B_z.

(x-components) $3.84 \times 10^{-19}\ \text{N} = (9.6 \times 10^{-16}\ \text{C·m/s})B_z \implies B_z = 4.0 \times 10^{-4}\ \text{T}.$

(y-components) $3.20 \times 10^{-19}\ \text{N} = (8.0 \times 10^{-16}\ \text{C·m/s})B_z \implies B_z = 4.0 \times 10^{-4}\ \text{T}$ (as before).

(z-components) $2.40 \times 10^{-19}\ \text{N} = (-8.0 \times 10^{-16}\ \text{C·m/s})B_y \implies B_y = -3.0 \times 10^{-4}\ \text{T}.$

Therefore, the magnetic field is

$$\vec{\mathbf{B}} = -(3.0 \times 10^{-4}\ \text{T})\hat{\mathbf{j}} + (4.0 \times 10^{-4}\ \text{T})\hat{\mathbf{k}}.$$

20.13

a)   Here's the picture.

b)   The magnetic force on the charge is $\vec{\mathbf{F}} = q\vec{\mathbf{v}} \times \vec{\mathbf{B}}$. So, the magnitude of the force is

$$F = |qvB\sin\theta| = |(-10.0 \times 10^{-6}\ \text{C})(2.00 \times 10^3\ \text{m/s})(15.0 \times 10^{-3}\ \text{T})\sin 75.0°| = 2.90 \times 10^{-4}\ \text{N}.$$

The particle is negatively charged, so the direction of the magnetic force is opposite that of $\vec{\mathbf{v}} \times \vec{\mathbf{B}}$. In the sketch above, the direction of $\vec{\mathbf{v}} \times \vec{\mathbf{B}}$ is into the page. Hence the magnetic force on the charge is perpendicular to and out of the page.

c)   According to Newton's second law

$$\vec{F} = m\vec{a} \implies \vec{a} = \frac{\vec{F}}{m} \implies a = \frac{2.90 \times 10^{-4} \text{ N}}{2.00 \times 10^{-3} \text{ kg}} = 0.145 \text{ m/s}^2 \,.$$

20.17

a)   The electric force on each charge is

$$\vec{F}_{\text{elec}} = q\vec{E}.$$

Since the particles are negatively charged, the direction of the electric force is opposite to that of the electric field. This is shown in the sketch below.

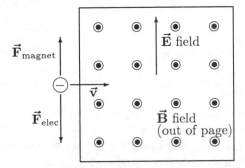

In a velocity selector, the direction of the magnetic force must be opposite to that of the electric force, in order for the two forces to sum to zero for the selected velocity. Hence the direction of the magnetic force on the particles must be as shown above.

b)   The magnetic force on a charge in a magnetic field is

$$\vec{F}_{\text{magnet}} = q\vec{v} \times \vec{B}.$$

In our case, q is negative, so the direction of $\vec{F}_{\text{magnet}}$ is opposite to the direction of $\vec{v} \times \vec{B}$. Hence we want $\vec{v} \times \vec{B}$ to be opposite to the direction of $\vec{F}_{\text{magnet}}$. Play with the vector product right-hand rule and you will see that we need $\vec{B}$ to point *out* of the page, as shown in the above sketch.

c)   For the selected morons, the electric and magnetic forces vector sum to zero, so their magnitudes must be equal. Hence, for the selected morons,

$$|q|E = |q|vB\sin\theta = |q|vB\sin 90° = |q|vB \implies v = \frac{E}{B} = \frac{2.00 \times 10^4 \text{ N/C}}{0.15 \text{ T}} = 1.3 \times 10^5 \text{ m/s} \,.$$

20.21   The magnetic force on a charge in a magnetic field is

$$\vec{F}_{\text{magnet}} = q\vec{v} \times \vec{B},$$

so the magnitude of this force on an electron is

$$F_{\text{magnet}} = evB\sin\theta.$$

The projection of the motion perpendicular to the field is a circle of radius r, along which the particle moves with speed $v\sin\theta$; the magnetic force also is perpendicular to the field and is the force that causes the centripetal acceleration. Apply Newton's second law to the particle in the plane of this circular motion; use the magnitudes of the vectors:

$$F_{\text{magnet}} = ma \implies evB\sin\theta = m\left(\frac{v^2\sin^2\theta}{r}\right) \implies r = \frac{mv\sin\theta}{eB}.$$

The diameter d of the circle is twice the radius, so

$$d = 2r = \frac{2mv \sin \theta}{eB}.$$

For small angles θ measured in radians , $\sin \theta \approx \theta$, so

$$d \approx \frac{2mv\theta}{eB}.$$

20.25

a)   The magnitude of the magnetic moment $\vec{\mu}$ of the loop is

$$\mu = IA = I\pi r^2 = (1.50 \text{ A })\pi(5.0 \times 10^{-2} \text{ m })^2 = 1.2 \times 10^{-2} \text{ A·m}^2$$

The direction of $\vec{\mu}$ is into the page.

b)   The magnitude of the torque on the loop is

$$\tau = |\vec{\mu} \times \vec{B}| = \mu B \sin \theta = (1.2 \times 10^{-2} \text{ A·m}^2)(0.60 \text{ T}) \sin 90° = 7.2 \times 10^{-3} \text{ N·m}.$$

c)   The direction of the torque is determined by applying the vector product right-hand rule to $\vec{\mu} \times \vec{B}$; the direction is indicated in the sketch below. Place the thumb of the right hand along the direction of the torque, then the curled fingers of the right hand indicate the sense that the loop will rotate about the line of the torque; this directional sense also is shown in the sketch below.

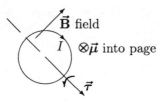

d)   Since $\vec{\mu}$ and $\vec{B}$ are perpendicular, the potential energy of the loop is

$$\text{PE} = -\vec{\mu} \bullet \vec{B} = \mu B \cos 90° = 0 \text{ J}.$$

20.29

a)   For purposes of computing the magnetic force on m, view it as a rigid wire carrying a current I in the direction indicated in Figure P.29. Then, from Equation 20.11 on page 905 of the text, the magnetic force $\vec{F}$ on m is

$$\vec{F}_{\text{magnet}} = I\vec{\ell} \times \vec{B},$$

where $\vec{\ell}$ is in the direction of the current. Since $\vec{\ell}$ and $\vec{B}$ are perpendicular, the magnitude of this force is

$$F = I\ell B \sin 90° = I\ell B.$$

b)   The forces on m are:

  1. its weight $\vec{w}$, of magnitude mg, directed straight down;

  2. the normal force $\vec{N}$ of the rails, directed perpendicular to the rails; and

  3. the magnetic force $\vec{F}_{\text{magnet}}$ of magnitude $I\ell B$, directed horizontally, as shown below.

Here's the second law force diagram together with a choice of coordinate system.

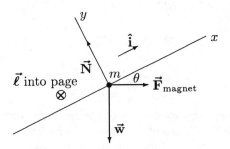

c) Apply Newton's second law in the $\hat{\mathbf{i}}$ direction, since the motion is in this direction.

$$F_{x\ \text{total}} = ma_x$$
$$\implies I\ell B\cos\theta - mg\sin\theta = ma_x$$

(1)
$$\implies a_x = \frac{I\ell B}{m}\cos\theta - g\sin\theta.$$

The acceleration component is constant, so we may apply the kinematic equation for motion with a constant acceleration. Assume that the bar starts from rest at time $t = 0$ s.

$$v_x(t) = v_{x0} + a_x t = a_x t = \left(\frac{I\ell B}{m}\cos\theta - g\sin\theta\right)t.$$

So

$$\vec{\mathbf{v}} = \left(\frac{I\ell B}{m}\cos\theta - g\sin\theta\right)t\hat{\mathbf{i}}.$$

d) The bar is in equilibrium if its acceleration is zero (which, in this case, also implies that its velocity is zero). Set the acceleration component given by equation (1) above equal to zero and solve for B.

$$\frac{I\ell B}{m}\cos\theta - g\sin\theta = 0 \text{ m/s}^2 \implies B = \frac{mg\sin\theta}{I\ell\cos\theta} = \frac{mg}{I\ell}\tan\theta.$$

20.33 The torque on the loop is $\vec{\tau} = \vec{\mu} \times \vec{\mathbf{B}}$, so the magnitude of the torque is

$$\tau = \mu B\sin\theta.$$

The maximum magnitude is when the angle θ between $\vec{\mu}$ and $\vec{\mathbf{B}}$ is $90°$, so

$$\tau = \mu B\sin\theta = \mu B.$$

The magnitude of the magnetic moment is the product of the current I, the area A of the coil, and the number n of coils,

$$\mu = nIA \implies \tau = nIAB \implies A = \frac{\tau}{nIB} = \frac{20.0 \text{ N·m}}{100(25.0 \text{ A})(0.200 \text{ T})} = 4.00 \times 10^{-2} \text{ m}^2.$$

20.37 We are asked to show that $\text{A·m}^2\text{·T} = \text{J}$.

Let's start with the left hand side. An ampere is defined as one coulomb per second,

$$\text{A} = \frac{\text{C}}{\text{s}},$$

and using the equation for the magnetic force, we see that a tesla is defined as one newton second per coulomb meter,

$$\text{T} = \frac{\text{N} \cdot \text{s}}{\text{C} \cdot \text{m}}.$$

Looking at the left hand side, we recall that a joule is defined as a newton meter,

$$\text{J} = \text{N} \cdot \text{m}.$$

Putting these definitions into our initial equation, we have

$$\text{A·m}^2\text{·T} = \frac{\text{C}}{\text{s}} \cdot \text{m}^2 \cdot \frac{\text{N} \cdot \text{s}}{\text{C} \cdot \text{m}} = \text{N} \cdot \text{m} = \text{J}.$$

20.41   Use the Biot-Savart law:

$$\vec{\mathbf{B}} = \frac{\mu_0}{4\pi} I \int \frac{d\vec{\ell} \times \hat{\mathbf{r}}}{r^2}$$

At every point along the straight segment to the left of the ring, the vector $d\vec{\ell}$ is parallel to $\hat{\mathbf{r}}$, so the vector product $d\vec{\ell} \times \hat{\mathbf{r}}$ vanishes. Therefore the current segments to the left of the ring contribute nothing to the magnetic field at point P. At every point along the straight segment to the right of the ring, the vector $d\vec{\ell}$ is antiparallel to $\hat{\mathbf{r}}$, so again the vector product $d\vec{\ell} \times \hat{\mathbf{r}}$ vanishes. Thus, neither of these portions of wire contribute to the magnetic field at point P. Whatever magnetic field is produced at P must be produced by the current in the ring.

Let $\hat{\mathbf{k}}$ be a unit vector pointing out of the page. First find the magnetic field created by the current through the *bottom* half of the ring. Just as in Example 20.9 on page 913 of the text, everywhere on the bottom half the angle between $d\vec{\ell}$ and $\hat{\mathbf{r}}$ is 90° and the right hand rule gives us $d\vec{\ell} \times \hat{\mathbf{r}} = d\ell\,\hat{\mathbf{k}}$. Thus

$$\vec{\mathbf{B}}_{\text{bottom}} = \frac{\mu_0}{4\pi} I \int_{\text{bottom}} \frac{d\vec{\ell} \times \hat{\mathbf{r}}}{r^2} = \frac{\mu_0}{4\pi} I \int_{\text{bottom}} \frac{d\ell}{r^2}\hat{\mathbf{k}} = \frac{\mu_0}{4\pi} I \frac{1}{R^2} \left(\int_{\text{bottom}} d\ell \right) \hat{\mathbf{k}}$$

Since $\int_{\text{bottom}} d\ell = \text{length of bottom} = \pi R$,

$$\vec{\mathbf{B}}_{\text{bottom}} = \frac{\mu_0}{4\pi} I \frac{1}{R^2} \pi R \hat{\mathbf{k}} = \frac{\mu_0 I}{4R} \hat{\mathbf{k}}.$$

Notice that this result is exactly one half of the result obtained in Example 20.9. That's reassuring!

Now find the magnetic field created by the current through the *top* of the ring. Just as for the bottom, everywhere on the top half the angle between $d\vec{\ell}$ and $\hat{\mathbf{r}}$ is 90°, but this time the right hand rule gives us $d\vec{\ell} \times \hat{\mathbf{r}} = d\ell(-\hat{\mathbf{k}})$. Thus, going through the same computation for the top of the ring,

$$\vec{\mathbf{B}}_{\text{top}} = -\frac{\mu_0 I}{4R} \hat{\mathbf{k}}.$$

The total magnetic field is the sum of these two fields and is therefore zero.

20.45

a)   Each side of the square produces a field at the center of the loop in the same direction. Hence the magnitude of the total field is 4 times the result of problem 20.44, with $z = \dfrac{\ell}{2}$ for the center of the square. Therefore

$$B = 4 \left(\frac{\mu_0}{4\pi} \frac{2I}{\dfrac{\ell}{2}} \frac{\ell}{\left(\ell^2 + 4\left(\dfrac{\ell}{2}\right)^2 \right)^{1/2}} \right) = \frac{\mu_0}{4\pi} 16 I \frac{1}{(2\ell^2)^{1/2}} = \frac{\mu_0}{4\pi} 8\sqrt{2} \frac{I}{\ell}$$

b)   Grab the loop with the fingers of your right hand with your thumb in the direction of the current around the loop. Your fingers indicate the sense in which the field passes through the loop. The direction of the field is perpendicular to and into the page.

20.49

a)   The magnitude of the magnetic field a distance d from an infinite wire is given by the result of Strategic Example 20.11 on pages 915-916 of the text:

$$B = \frac{\mu_0}{4\pi} \frac{2I}{d} = (10^{-7}\ \text{T·m/A}) \frac{2(15\ \text{A})}{1.00 \times 10^{-2}\ \text{m}} = 3.0 \times 10^{-4}\ \text{T}.$$

The magnetic field at the position of the proton is directed into the page.

b)   The force on a charged particle moving in a magnetic field is

$$\vec{F} = q\vec{v} \times \vec{B}.$$

The magnitude of this force is

$$F = |q|vB\sin\theta = (1.602 \times 10^{-19}\ \text{C})(5.0 \times 10^{7}\ \text{m/s})(3.0 \times 10^{-4}\ \text{T})\sin 90° = 2.4 \times 10^{-15}\ \text{N}.$$

Since q is positive, $\vec{F}$ is in the same direction as $\vec{v} \times \vec{B}$. By the vector product right-hand rule, the direction of $\vec{F} = q\vec{v} \times \vec{B}$ is as shown in the sketch below.

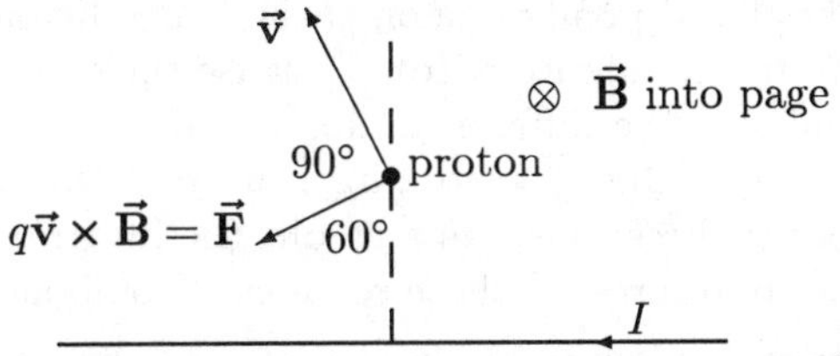

c)   The force of the wire on the proton and the force of the proton on the wire form a Newton's third law force pair. Therefore, the force of the proton on the wire has magnitude 2.4×10^{-15} N and is directed opposite to the force $\vec{F}$ shown in the sketch above.

d)   The magnetic force does no work. Hence, from the CWE theorem, the speed of the proton is unchanged. It remains 5.0×10^{7} m/s.

20.53

a)   The magnitude of the magnetic field a distance r from an infinite wire is given by the result of Strategic Example 20.11 on pages 915-916 of the text. At the most distant segment of the rectangle, the field of the infinite wire has a magnitude

$$B = \frac{\mu_0}{4\pi}\frac{2I}{r} = (10^{-7}\ \text{T·m/A})\frac{2(15\ \text{A})}{40 \times 10^{-2}\ \text{m}} = 7.5 \times 10^{-6}\ \text{T}.$$

The direction of the field is found by grasping the wire causing the field with your right hand with the thumb in the direction of the current; the fingers then wrap around the wire in the sense of the direction of the magnetic field. The magnetic field of the infinite wire at the position of the side of the rectangular loop most distant from the wire is perpendicular to and out of the page.

b)   The force on a straight, current carrying wire in a magnetic field that is constant along its length is

$$\vec{F}_{\text{distant}} = I\vec{\ell} \times \vec{B}.$$

Here $\vec{\ell}$ and $\vec{B}$ are perpendicular to each other so

$$F_{\text{distant}} = I\ell B = (5.0\ \text{A})(0.50\ \text{m})(7.5 \times 10^{-6}\ \text{T}) = 1.9 \times 10^{-5}\ \text{N}.$$

Use the vector product right-hand rule to determine the direction of $\vec{F}_{\text{distant}}$. The force is *toward* the infinite wire.

c)   The magnitude of the magnetic field a distance r from an infinite wire is given by the result of Strategic Example 20.11. At the closest segment of the wire, the field of the infinite wire has a magnitude

$$B = \frac{\mu_0}{4\pi}\frac{2I}{r} = (10^{-7}\ \text{T·m/A})\frac{2(15\ \text{A})}{20 \times 10^{-2}\ \text{m}} = 15 \times 10^{-6}\ \text{T}.$$

The direction of the field is found by grasping the wire causing the field with your right hand with the thumb in the direction of the current; the fingers then wrap around the wire in the sense of the direction of the magnetic field. The magnetic field of the infinite wire at the position of the side of the rectangular loop closest to the wire is perpendicular to and out of the page.

d)   The force on a straight, current carrying wire in a magnetic field that is constant along its length is

$$\vec{\mathbf{F}}_{\text{close}} = I\vec{\ell} \times \vec{\mathbf{B}}.$$

Here $\vec{\ell}$ and $\vec{\mathbf{B}}$ are perpendicular to each other so

$$F_{\text{close}} = I\ell B = (5.0 \text{ A})(0.50 \text{ m})(15 \times 10^{-6} \text{ T}) = 3.8 \times 10^{-5} \text{ N}.$$

Use the vector product right-hand rule to determine the direction of $\vec{\mathbf{F}}_{\text{close}}$. Its direction is *away* from the infinite wire.

e)   Consider two symmetrically placed points one on each perpendicular segment of the rectangle (that is, two points the same distance from the infinite wire). Because they are the same distance from the infinite wire, the (differential) magnitudes of the differential force will be the same at these two points. But the current in the left segment is going in the opposite direction from the current in the right segment. Hence, the differential forces have opposite directions. As a result, the integral of the differential forces on the two sides cancel each other, and the total force on these segments (combined) is zero.

f)   The total force on the rectangular loop is the vector sum of the forces on the four sides. The forces on the left and right sides vector sum to zero. The force on the near segment is away from the infinite wire while the force on the more distant side is toward the infinite wire. The vector sum of the forces has a magnitude equal to

$$3.8 \times 10^{-5} \text{ N} - 1.9 \times 10^{-5} = 1.9 \times 10^{-5} \text{ N}.$$

Thus, the total force on the rectangular loop has a magnitude of 1.9×10^{-5} N and is directed *away* from the infinite wire.

g)   The force of the infinite wire on the rectangular loop and the force of the rectangular loop on the infinite wire form a Newton's third law force pair. Hence the force of the rectangular loop on the infinite wire has a magnitude of 1.9×10^{-5} N and is directed *away* from the loop.

20.57   The Biot-Savart law implies that the current in the cylindrical sheath produces no magnetic field along its axis. To see this, imagine two slices of the outside sheath taken along its length. The two slices are on opposite sides of the cable, but otherwise have the same dimensions, in particular they each carry the same proportion of the current. Here's a cross-sectional picture of the situation.

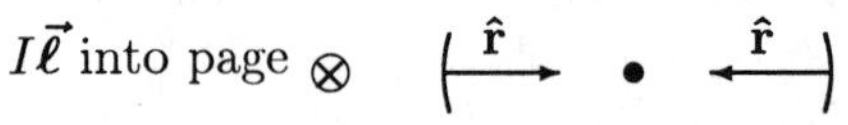

In the sketch we have supposed the current is going into the page. By symmetry, the magnetic fields produced by each of the two slices have the same *magnitudes* at the center. But, using the right hand rule, the magnetic field produced by the left hand current points straight down this page, while the field produced by the right half points straight up this page. Hence, they vector sum to zero. Thus the vector field produced by any slice of the outer sheath is exactly canceled by the field produce by a corresponding slice on the opposite side, so the total magnetic field is zero.

a)   The problem asks for the field very close to the axis of the cable. Since the field produced by the outer cable is zero at the axis, we may neglect it and concentrate on the field produced by the central conductor. This conductor produces a field like that of an infinite wire, whose magnitude a perpendicular distance r from the wire is

$$B = \frac{\mu_0}{4\pi}\frac{2I}{r} = (10^{-7} \text{ T·m/A})\frac{2(3.0 \text{ A})}{\left(\dfrac{1.0 \times 10^{-3} \text{ m}}{2}\right)} = 1.2 \times 10^{-3} \text{ T}.$$

The direction of the field is found by grasping the wire causing the field with your right hand with the thumb in the direction of the current; the fingers then wrap around the wire in the sense of the direction of the magnetic field. The direction of the field is shown below.

b) Outside the coaxial cable, the magnetic field is the vector sum of the fields produced by the inside wire and the outside sheath. Since the currents have the same magnitude, but are in opposite directions, the vector sum of the their magnetic fields rapidly falls to zero as one moves away form the cable. (Because the magnitude of the field falls off as $\frac{1}{r}$, the field produced by the outside conductor is slightly greater in magnitude than the field produced by the inside conductor, but this effect rapidly becomes negligible.)

20.61

a)   We'll generalize this a tiny bit. Instead of the wire going from the origin to the point P, let the wire go from an arbitrary point P' to the point P.

The force on a current carrying wire in a magnetic field is

$$\vec{\mathbf{F}}_{\text{wire}} = I \int_{\text{wire}} d\vec{\boldsymbol{\ell}} \times \vec{\mathbf{B}}.$$

The magnetic field is $\vec{\mathbf{B}} = B\hat{\mathbf{k}}$. Write $d\vec{\boldsymbol{\ell}} = dx\hat{\mathbf{i}} + dy\hat{\mathbf{j}}$. Then

$$\vec{\mathbf{F}}_{\text{wire}} = I \int_{\text{wire}} \left(dx\hat{\mathbf{i}} + dy\hat{\mathbf{j}} \right) \times B\hat{\mathbf{k}} = I \int_{\text{wire}} \left(B\,dx(-\hat{\mathbf{j}}) + B\,dy\hat{\mathbf{i}} \right)$$

The magnetic field is uniform, so we may bring B outside the integral. Let (x', y') be the coordinates of the point P', and let (x, y) be the coordinates of P. Then

$$\vec{\mathbf{F}}_{\text{wire}} = IB \int_{\text{wire}} \left(dx(-\hat{\mathbf{j}}) + dy\hat{\mathbf{i}} \right) = \left(IB \int_{x'}^{x} dx \right) (-\hat{\mathbf{j}}) + \left(IB \int_{y'}^{y} dy \right) \hat{\mathbf{i}} = IB(y - y')\hat{\mathbf{i}} - IB(x - x')\hat{\mathbf{j}}.$$

Notice that for the given constant magnetic field $\vec{\mathbf{B}}$ and given current I, the force $\vec{\mathbf{F}}_{\text{wire}}$ depends only on the end points of the wire — P' and P. It doesn't matter what sort of crazy path the wire may have taken to get from P' to P. In particular, $\vec{\mathbf{F}}_{\text{wire}}$ would have been the same if the wire had gone straight from P' to P.

b)   If the wire is a loop, so that $P = P'$, then $x - x' = 0$ m and $y - y' = 0$ m, so $\vec{\mathbf{F}}_{\text{wire}} = \mathbf{0}$ N .

20.65

a)   The force on the electron produced by the magnetic field $\vec{\mathbf{B}}$ is $\vec{\mathbf{F}} = (-e)\vec{\mathbf{v}} \times \vec{\mathbf{B}}$. By definition of the vector product, this vector is perpendicular to $\vec{\mathbf{v}}$. Therefore, the force is pointing either directly towards the center of the circle, or directly away from it. Since the electron is accelerating towards the center, the force must be pointed towards the center rather than away from it.

b)   The force produced by the magnetic field $\vec{\mathbf{B}}$ on the electron is

$$\vec{\mathbf{F}} = (-e)\vec{\mathbf{v}} \times \vec{\mathbf{B}}.$$

This is opposite in direction to $\vec{\mathbf{v}} \times \vec{\mathbf{B}}$. Thus, we need $\vec{\mathbf{v}} \times \vec{\mathbf{B}}$ to be directed *away* from the center of the circle. Play around with the right hand rule and you'll see that this means that $\vec{\mathbf{B}}$ must be directed perpendicular to and into the page. Here's the picture.

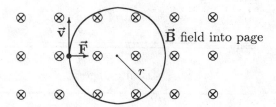

c) Use magnitudes in Newton's second law,

$$F = ma \implies |-e|vB \sin 90° = m\frac{v^2}{r} \implies r = \frac{mv}{eB}.$$

d) Since the magnetic force is perpendicular to the change in the position vector at all points along the path, the work done by the magnetic force is 0 J.

e) The force on the proton is

$$\vec{\mathbf{F}} = e\vec{\mathbf{v}} \times \vec{\mathbf{B}}.$$

Since the proton is at rest, then $\vec{\mathbf{v}} = \mathbf{0}$ m/s, and therefore $\vec{\mathbf{F}} = \mathbf{0}$ N. The proton stays at rest.

20.69

a) The force on the electron is

$$\vec{\mathbf{F}} = (-e)\vec{\mathbf{v}} \times \vec{\mathbf{B}}.$$

The magnitude of this force is

$$F = evB \sin \theta,$$

where θ is the angle between the magnetic field $\vec{\mathbf{B}}$ and the velocity $\vec{\mathbf{v}}$. For fixed v and B, this magnitude is a maximum when $\theta = 90°$. In this case,

$$F_{\text{max}} = evB \sin 90° = evB = (1.602 \times 10^{-19}\ \text{C})(7.50 \times 10^6\ \text{m/s})(50.0 \times 10^{-3}\ \text{T}) = 6.00 \times 10^{-14}\ \text{N}.$$

b) In this case

$$F = 0.25 F_{\text{max}} \implies evB \sin \theta = 0.25 evB \implies \sin \theta = 0.25 \implies \theta = 14°.$$

c) The initial kinetic energy is

$$\text{KE} = \frac{1}{2}mv^2 = \frac{1}{2}(9.11 \times 10^{-31}\ \text{kg})(7.50 \times 10^6\ \text{m/s})^2 = 2.56 \times 10^{-17}\ \text{J}$$

$$= (2.56 \times 10^{-17}\ \text{J})\left(\frac{\text{eV}}{1.602 \times 10^{-19}\ \text{J}}\right) = 160\ \text{eV}.$$

d) The magnetic force does no work. Hence, by the CWE theorem, the kinetic energy stays at 160 eV.

20.73 The closed path has no currents piercing it, so according to Ampere's law the integral of $\vec{\mathbf{B}} \bullet d\vec{\boldsymbol{\ell}}$ around the closed path should be zero. Let's check it out.

Integrating in a counter-clockwise direction, beginning with the bottom, we have

$$\int_{\text{clsd path}} \vec{\mathbf{B}} \bullet d\vec{\boldsymbol{\ell}} = \int_{\text{bottom}} \vec{\mathbf{B}} \bullet d\vec{\boldsymbol{\ell}} + \int_{\text{right side}} \vec{\mathbf{B}} \bullet d\vec{\boldsymbol{\ell}} + \int_{\text{top}} \vec{\mathbf{B}} \bullet d\vec{\boldsymbol{\ell}} + \int_{\text{left side}} \vec{\mathbf{B}} \bullet d\vec{\boldsymbol{\ell}}$$

Along the top and bottom, $\vec{\mathbf{B}}$ and $d\vec{\ell}$ are perpendicular, so the corresponding integrals are zero. Along the right side, $\vec{\mathbf{B}}$ is zero, so that integral is also zero. We're left with

$$\int_{\text{clsd path}} \vec{\mathbf{B}} \bullet d\vec{\ell} = \int_{\text{left side}} \vec{\mathbf{B}} \bullet d\vec{\ell}$$

Along the left side, $d\vec{\ell}$ and $\vec{\mathbf{B}}$ are parallel, so $\vec{\mathbf{B}} \bullet d\vec{\ell} = B d\ell$. Therefore

$$\int_{\text{clsd path}} \vec{\mathbf{B}} \bullet d\vec{\ell} = \int_{\text{left side}} B \, d\ell = B \int_{\text{left side}} d\ell = B(\text{length of left side}) = B\ell.$$

Unless $\vec{\mathbf{B}}$ is zero, $B\ell$ is *not* zero. This contradiction of Ampere's law shows that the supposed vector field is physically impossible.

There must be some sort of turning of the vector field as one gets close to its edge (in order to allow the top and bottom integrals to be non zero), or the vector field on the right side of the picture must also be nonzero.

Chapter 21

Faraday's Law of Electromagnetic Induction

21.1 From its definition, page 898 of the text, one tesla is $T = \dfrac{N}{C \cdot m/s}$. Also, one newton·meter is a joule, $N \cdot m = J$, and one joule per coulomb is a volt, $J/C = V$. Hence,

$$T \cdot m^2/s = \frac{N}{C \cdot m/s} \cdot m^2/s = \frac{N \cdot m}{C} = \frac{J}{C} = V.$$

21.5

a) Since the maximum value of the sine is 1, the maximum value of the output is $311\,V$.

b) The angular frequency ω of the source is the coefficient of t in the argument of the sine. Also, the angular frequency ω and frequency ν are related by $\omega = 2\pi\nu$. Hence

$$\omega = 314\,\text{rad/s} \implies 2\pi\nu = 314\,\text{rad/s} \implies \nu = 50.0\,\text{Hz}.$$

21.9

a) The charge separation produces an electric field that produces an electric force on the charges in a direction opposite to the direction of the magnetic force on them. The magnitude of the electric force eventually becomes equal to (approaches) the magnitude of the magnetic force. After that no further charge separation occurs.

b) From the vector product right-hand rule, the direction of $\vec{v} \times \vec{B}$ is along the rod and upward. Therefore, positive charge accumulates on the upper end of the rod and negative charge at the lower end.

c) When the magnitude of the electric force on a charge q is equal to that of the magnetic force, we have

$$qE = qvB \implies E = vB.$$

The absolute value of the potential difference V between the ends of the rod is

$$|V| = E\ell = vB\ell.$$

d) The rod moves a distance $v\Delta t$ during the time Δt, so the area swept out by the rod is

$$\ell v \Delta t.$$

e)

$$\left| \frac{\Delta \Phi}{\Delta t} \right| = \frac{B\ell v \Delta t}{\Delta t} = B\ell v.$$

This is the same as the result of part c).

This type of charge separation, and the resulting electric field, can actually occur on airplanes as they fly through the Earth's magnetic field.

21.13

a) Since the magnetic field is the same over the whole circular area, the magnetic flux through the path is

$$\Phi = \vec{B} \bullet \vec{A},$$

where $\vec{A}$ is the area vector. Hence

$$\Phi = BA\cos\theta = B_0 e^{-\alpha t}\pi r^2 \cos\theta.$$

b) The induced emf around the path is

$$\text{induced emf} = -\frac{d\Phi}{dt} = \alpha B_0 e^{-\alpha t}\pi r^2 \cos\theta.$$

Since the magnetic field is decreasing with time, the flux through the path also is decreasing with time. Lenz's law states that the induced current is directed to oppose the change, so the magnetic field produced by the induced current will be in the *same* direction as the given field through the path. Grasp the loop with the fingers on your right hand threading the loop along $\hat{\mathbf{k}}$. The right hand thumb indicates the direction of the induced current around the loop. Here's the picture.

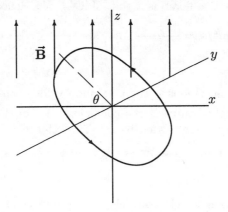

21.17

a) The induced emf is

$$\text{induced emf} = -\frac{d\Phi}{dt} = -B\ell v = -(0.5 \times 10^{-4}\ \text{T})(1.36\ \text{m})(50.0\ \text{m/s}) = -3 \times 10^{-3}\ \text{V}.$$

b) The component of the magnetic field that contributes to the time rate of change of the magnetic flux must be perpendicular to both the axle of the train and to the velocity vector of the axle. This is the vertical component of the magnetic field.

c) The magnetic force $\vec{F} = q\vec{v} \times \vec{B}$ on the mobile charge carriers inside a conductor leads to charge separation and an electric field between the separated charges.

21.21

a) The frequency ν is 50.0 Hz, so the angular frequency is

$$\omega = 2\pi\nu = 2\pi(50.0\ \text{Hz}) = 314\ \text{rad/s}.$$

The coil should be turned at this angular frequency.

b) Use Equation 21.8 on page 964 of the text:

$$\text{induced emf} = NBA\omega \sin(\omega t),$$

where $N = 300$ is the number of turns, B is the (unknown) magnitude of the field, and $A = (10 \times 10^{-2} \text{ m})(20 \times 10^{-2} \text{ m})$ is the area of each loop. The amplitude is the peak induced emf. This occurs when $\sin(\omega t) = 1$. Thus

$$\text{peak induced emf} = NBA\omega$$

$$\Longrightarrow B = \frac{\text{peak induced emf}}{NA\omega} = \frac{12.0 \text{ V}}{300(10 \times 10^{-2} \text{ m})(20 \times 10^{-2} \text{ m})(314 \text{ rad/s})} = 6.4 \times 10^{-3} \text{ T}.$$

21.25 The favorite radio station of one of the authors has a frequency of $\nu = 89.1$ MHz. Its wavelength is

$$\lambda = \frac{c}{\nu} = \frac{3.00 \times 10^8 \text{ m/s}}{89.1 \times 10^6 \text{ Hz}} = 3.37 \text{ m}.$$

21.29 Use Equation 21.34 on page 971 of the text:

$$V = L\frac{dI}{dt}.$$

a)

$$V = (150 \times 10^{-3} \text{ H})(60 \text{ A/s}) = 9.0 \text{ V}.$$

b)

$$V = (150 \times 10^{-3} \text{ H})(-50 \text{ A/s}) = -7.5 \text{ V}.$$

c)

$$V = (150 \times 10^{-3} \text{ H})(40 \text{ A/s}) = 6.0 \text{ V}.$$

21.33

a) We'll first show that the magnetic field inside the outer sheath created by current in the outer sheath is zero, so then we can concentrate solely on the field created by the current in the central conductor.

To see this, let's first ignore the inner conductor and consider a point P a distance D from the center of the outer sheath. At that point, let $\hat{\ell}$ point along the cable in the direction of the outer conductor current, let $\hat{r}$ be a unit vector perpendicular to $\hat{\ell}$ pointing from P towards the center of the conductor, and let $\hat{\theta}$, be a unit vector perpendicular to both $\hat{\ell}$ and $\hat{r}$ chosen so that $\hat{r}$, $\hat{\theta}$, $\hat{\ell}$ form a right handed system. Here's the picture.

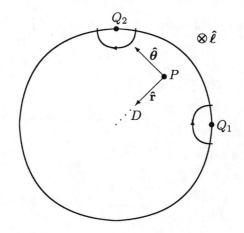

Let $\vec{\mathbf{B}}$ be the magnetic field produced by the outside conductor, and write the vector $\vec{\mathbf{B}}$ at P in the form

$$\vec{\mathbf{B}} = B_r \hat{\mathbf{r}} + B_\theta \hat{\boldsymbol{\theta}} + B_\ell \hat{\boldsymbol{\ell}}.$$

From the Biot-Savart law, $\vec{\mathbf{B}}$ and $\hat{\boldsymbol{\ell}}$ are perpendicular, hence B_ℓ is zero. For each point Q_1 on the outer conductor there is another point Q_2 symmetrically located with respect to P whose differential contribution to the B_r term at P is equal in magnitude but *opposite* in direction to the differential contribution made by Q_1. Therefore the total contribution to the B_r term by all the points Q around the outer sheath is zero. Thus we are left with

$$\vec{\mathbf{B}} = B_\theta \hat{\boldsymbol{\theta}}.$$

Now let C be the concentric circle of radius D passing through P, and integrate $\vec{\mathbf{B}} \bullet d\vec{\mathbf{r}}$ in a counter-clockwise direction around C. Then around that circle we have $\vec{\mathbf{B}} \bullet d\vec{\mathbf{r}} = B_\theta D\, d\theta$. Also, by symmetry, B_θ depends only on the distance D of P from the center and is therefore constant all along the path C. Hence

$$\int_C \vec{\mathbf{B}} \bullet d\vec{\mathbf{r}} = \int_0^{2\pi} B_\theta D\, d\theta = B_\theta D \int_0^{2\pi} d\theta = 2\pi D B_\theta.$$

From Ampere's law $\int_C \vec{\mathbf{B}} \bullet d\vec{\mathbf{r}} = \mu_0 I_{\text{net current threading the path}}$. Since there is no current threading the path, the integral is zero. Thus

$$\int_C \vec{\mathbf{B}} \bullet d\vec{\mathbf{r}} = 0 \text{ T}\cdot\text{m} \implies 2\pi D B_\theta = 0 \text{ T}\cdot\text{m} \implies \vec{\mathbf{B}} = B_\theta \hat{\boldsymbol{\theta}} = \mathbf{0}\text{ T}.$$

Thus the magnetic field inside the coaxial cable is produced entirely by the inner conductor.

Now let $\vec{\mathbf{B}}$ be the magnetic field produced by the inner conductor. As we've just seen, this is the total magnetic field inside the outer sheath.

The magnetic field at a distance r from an infinitely long wire is

$$\vec{\mathbf{B}} = \frac{\mu_0}{4\pi} \frac{2I}{r} \hat{\boldsymbol{\theta}}.$$

Here I is the magnitude of the current, and $\hat{\boldsymbol{\theta}} = \hat{\boldsymbol{\ell}} \times \hat{\mathbf{r}}$, where $\hat{\boldsymbol{\ell}}$ is a unit vector pointing along the wire in the direction of the current and $\hat{\mathbf{r}}$ is a unit vector perpendicular to $\hat{\boldsymbol{\ell}}$ pointing *from* the wire *to* the point.

Consider the differential flux $d\Phi$ through a differential strip of the rectangle described in the problem. Here's the picture.

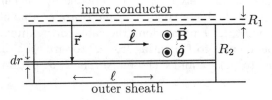

The differential strip has length ℓ and width dr. The differential flux through this strip is

$$d\Phi = \vec{\mathbf{B}} \bullet d\vec{\mathbf{A}} = \frac{\mu_0}{4\pi} \frac{2I}{r} \ell\, dr.$$

Hence the total flux through the rectangle is

$$\Phi = \int_{R_1}^{R_2} \vec{\mathbf{B}} \bullet d\vec{\mathbf{A}} = \int_{R_1}^{R_2} \frac{\mu_0}{4\pi} \frac{2I}{r} \ell\, dr = \frac{\mu_0}{4\pi} 2I\ell(\ln R_2 - \ln R_1) = \frac{\mu_0}{4\pi} 2I\ell \ln\left(\frac{R_2}{R_1}\right).$$

b)   From part a), the self inductance per unit length is

$$\frac{L}{\text{length}} = \frac{\Phi}{I\ell} = \frac{\dfrac{\mu_0}{4\pi} 2I\ell \ln\left(\dfrac{R_2}{R_1}\right)}{I\ell} = \frac{\mu_0}{4\pi} 2 \ln\left(\frac{R_2}{R_1}\right).$$

c) From part b),

$$\frac{L}{\text{length}} = \frac{\mu_0}{4\pi} 2\ln\left(\frac{R_2}{R_1}\right) = (10^{-7} \text{ T·m/A})2\ln\left(\frac{1.7 \times 10^{-3} \text{ m}}{0.41 \times 10^{-3} \text{ m}}\right) = 2.8 \times 10^{-7} \text{ H/m}.$$

21.37

a) and

b) Since $V = L\dfrac{dI}{dt}$, then $\dfrac{dI}{dt} = \dfrac{V}{L}$. So, integrating this,

$$I(t) = I_0 + \frac{1}{L}\int_{t_0}^{t} V \, dt.$$

During the interval $0 \text{ s} \leq t \leq 2.0 \times 10^{-3} \text{ s}$, we have $I_0 = 0 \text{ A}$, and, from the graph in the problem statement, $V = \left(\dfrac{5.0 \text{ V}}{2.0 \times 10^{-3} \text{ s}}\right)t$. Therefore

$$I(t) = 0 \text{ A} + \frac{1}{L}\int_{0\text{ s}}^{t}\left(\frac{5.0 \text{ V}}{2.0 \times 10^{-3} \text{ s}}\right)t\,dt = \frac{1}{L}\left(\frac{5.0 \text{ V}}{2.0 \times 10^{-3} \text{ s}}\right)\frac{t^2}{2}$$

$$= \frac{1}{150 \times 10^{-3} \text{ H}}\left(\frac{5.0 \text{ V}}{2.0 \times 10^{-3} \text{ s}}\right)\frac{t^2}{2} = (8.3 \times 10^3 \text{ A/s}^2)t^2.$$

So, at time $t = 2.0 \times 10^{-3} \text{ s}$,

$$I(2.0 \times 10^{-3} \text{ s}) = (8.3 \times 10^3 \text{ A/s}^2)(2.0 \times 10^{-3} \text{ s})^2 = 33 \times 10^{-3} \text{ A}.$$

Thus, for the next interval of time, $2.0 \times 10^{-3} \text{ s} \leq t \leq 4.0 \times 10^{-3} \text{ s}$, we have $I_0 = 33 \times 10^{-3} \text{ A}$, and, from the graph in the problem statement, $V = 10.0 \text{ V} - \left(\dfrac{5.0 \text{ V}}{2.0 \times 10^{-3} \text{ s}}\right)t$. Therefore during this interval

$$I(t) = 33 \times 10^{-3} \text{ A} + \frac{1}{L}\int_{2.0\times10^{-3}\text{ s}}^{t}\left(10.0 \text{ V} - \frac{5.0 \text{ V}}{2.0 \times 10^{-3} \text{ s}}\right)t\,dt$$

$$= 33 \times 10^{-3} \text{ A} + \frac{1}{L}\left((10.0 \text{ V})t - \frac{5.0 \text{ V}}{2.0 \times 10^{-3} \text{ s}}\frac{t^2}{2}\right)\Bigg|_{2.0\times10^{-3}\text{ s}}^{t}$$

$$= -0.067 \text{ A} + (67 \text{ A/s})t - (8.3 \times 10^3 \text{ A/s}^2)t^2.$$

Thus, summarizing,

$$I(t) = \begin{cases} (8.3 \times 10^3 \text{ A/s}^2)t^2 & \text{for } 0 \text{ s} \leq t \leq 2.0 \times 10^{-3} \text{ s, and} \\ -0.067 \text{ A} + (67 \text{ A/s})t - (8.3 \times 10^3 \text{ A/s}^2)t^2 & \text{for } 2.0 \times 10^{-3} \text{ s} \leq t \leq 4.0 \times 10^{-3} \text{ s.} \end{cases}$$

Here's the graph of current as a function of time.

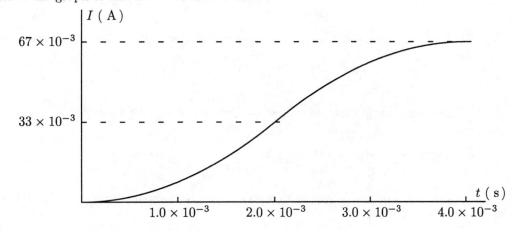

c)   The current is $1/L$ times the area under the $V(t)$ versus t graph in the text. For example, when $t = 4.0 \times 10^{-3}$ s, the current is

$$\frac{1}{L}\left(\frac{(\text{base})(\text{height})}{2}\right) = \frac{1}{150 \times 10^{-3}\text{ H}}\left(\frac{(4.0 \times 10^{-3}\text{ s})(5.0\text{ V})}{2}\right) = 67 \times 10^{-3}\text{ A}.$$

21.41   Use Equation 21.45 on page 975 of the text:

$$I(t) = \frac{V_0}{R}\left(1 - e^{-(R/L)t}\right) = \frac{V_0}{R}\left(1 - e^{-t/\tau}\right).$$

The final value of the current is $\dfrac{V_0}{R}$, so when the current is 0.9944 of the final value, then

$$0.9944\frac{V_0}{R} = \frac{V_0}{R}\left(1 - e^{-t/\tau}\right) \implies e^{-t/\tau} = .0056 \implies -\frac{t}{\tau} = \ln 0.0056 = -5.2 \implies t = 5.2\tau.$$

So, when t is 5.2 time constants, then the current has reached 99.44% of its final value.

21.45

a)   The time constant of the circuit is

$$\tau = \frac{L}{R} = \frac{250.0 \times 10^{-3}\text{ H}}{150\ \Omega} = 1.67 \times 10^{-3}\text{ s}.$$

b)   In the steady state, the inductor acts as a short circuit. The steady state current is

$$I_{\text{steady state}} = \frac{V_0}{R} = \frac{12.0\text{ V}}{150\ \Omega} = 80.0 \times 10^{-3}\text{ A}.$$

c)   Use Equation 21.45 on page 975 of the text:

$$I(t) = \frac{V_0}{R}\left(1 - e^{-(R/L)t}\right) = \frac{V_0}{R}\left(1 - e^{-t/\tau}\right).$$

The final value of the current is $\dfrac{V_0}{R}$, so when the current is 50% of this value, we have

$$0.50\frac{V_0}{R} = \frac{V_0}{R}\left(1 - e^{-t/\tau}\right) \implies e^{-t/\tau} = 0.50 \implies -\frac{t}{\tau} = \ln 0.50$$

$$\implies t = -\tau \ln 0.50 = -(1.67 \times 10^{-3}\text{ s})\ln 0.50 = 1.2 \times 10^{-3}\text{ s}.$$

So, after 1.2 milliseconds the current has attained 50% of its steady state value.

21.49

a)   According to Ohm's law, the current through a resistor and the potential difference across it are proportional to each other:

$$V = IR$$

The graphs indicate that I is not proportional to V, so the device cannot be a resistor.

b)   For a capacitor,

$$C = \frac{Q}{V} \implies Q = CV \implies I = \frac{dQ}{dt} = C\frac{dV}{dt}.$$

But the graphs indicate that I is not proportional to the slope of the V versus t graph, so the device is not a capacitor.

c)   and

d)   For an inductor

$$V = L\frac{dI}{dt}.$$

The slope of the I versus t graph is constant over the same intervals that V is constant, is negative over the same intervals that V is negative, and is positive over the same intervals that V is positive. It appears that the graphs are consistent with the device being an inductor. Looking more closely, we have,

time interval	$\frac{dI}{dt}$	V	$\frac{V}{\left(\frac{dI}{dt}\right)}$
0.00 s $< t <$ 0.50 s	−20 A/s	−2.0 V	0.10 H
0.50 s $< t <$ 1.5 s	10 A/s	1.0 V	0.10 H
1.50 s $< t <$ 2.0 s	0 A/s	0 V	undefined
2.0 s $< t <$ 3.0 s	−20 A/s	?	?
3.0 s $< t <$ 4.5 s	0 A/s	?	?

the data are consistent with the device being a 0.10 H inductor.

e)   Complete the above table letting $V = L\frac{dI}{dt}$ with $L = 0.10$ H:

time interval	$\frac{dI}{dt}$	V	$\frac{V}{\left(\frac{dI}{dt}\right)}$
0.00 s $< t <$ 0.50 s	−20 A/s	−2.0 V	0.10 H
0.50 s $< t <$ 1.5) s	10 A/s	1.0 V	0.10 H
1.50 s $< t <$ 2.0 s	0 A/s	0 V	undefined
2.0 s $< t <$ 3.0 s	−20 A/s	−2.0 V	0.10 H
3.0 s $< t <$ 4.5 s	0 A/s	0 V	0.10 H

Now use the completed table to complete the graph of V versus t:

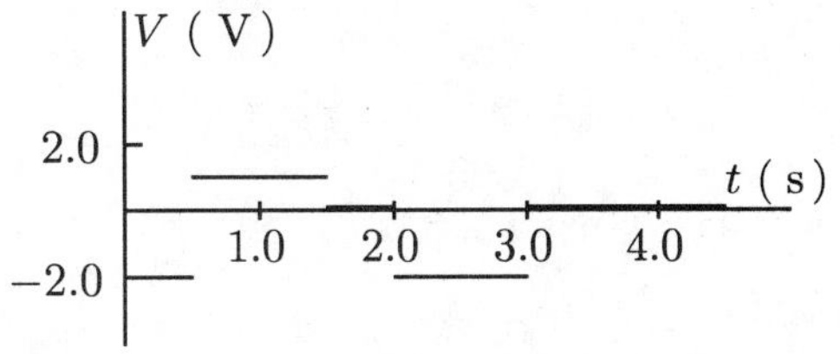

21.53   The angular frequency of the oscillation is

$$\omega = \frac{1}{\sqrt{LC}}.$$

The frequency ν is

$$\nu = \frac{\omega}{2\pi} = \frac{\frac{1}{\sqrt{LC}}}{2\pi} = \frac{1}{2\pi\sqrt{LC}} \implies C = \frac{1}{4\pi^2 L\nu^2}$$

For the low frequency end of the AM band, we need

$$C = \frac{1}{4\pi^2 (0.33 \times 10^{-3} \text{ H})(500 \times 10^3 \text{ Hz})^2} = 3.1 \times 10^{-10} \text{ F} = 3.1 \times 10^2 \text{ pF}.$$

For the high frequency end of the AM band, we need

$$C = \frac{1}{4\pi^2 (0.33 \times 10^{-3} \text{ H})(1600 \times 10^3 \text{ Hz})^2} = 3.0 \times 10^{-11} \text{ F} = 30 \text{ pF}.$$

Hence the capacitor must be able to vary its capacitance over the range from 30 pF to 3.1×10^2 pF.

21.57   Use Equation 21.66 on page 986 of the text:

$$\frac{N_1}{N_2} = \frac{V_1}{V_2} = \frac{14 \times 10^3 \text{ V}}{220 \text{ V}} = 64.$$

21.61

a)   and

b)   The emf and potential difference V have opposite polarity.

   For sketch (a), the magnetic field of the primary coil is increasing because of the increasing current. The magnetic field of the primary is directed through the secondary in the top to bottom direction. The flux of the primary through the secondary is increasing with time; Lenz's law opposes this change. In order to create the opposing magnetic field, the induced current must flow from terminal (4) to terminal (3). Hence terminal (4) of the secondary has the positive polarity for the emf and terminal (3) has the positive polarity for the potential difference.

   For sketch (b), the magnetic field of the primary coil is increasing because of the increasing current. The magnetic field of the primary is directed through the secondary in the left-to-right direction. The flux of the primary through the secondary is increasing with time; Lenz's law opposes this change. In order to create the opposing magnetic field, the induced current must flow from terminal (3) to terminal (4). Hence terminal (3) has the positive polarity for the emf and terminal (4) has the positive polarity for the potential difference.

   Note that the induced emf is oriented so that the induced current flows from its positive side to its negative side. On the other hand, the potential difference is oriented so that it is the negative of the induced emf; i.e., so that current flowing into the inductor from outside enters the terminal with the higher potential and leaves from the terminal with the lower potential.

Chapter 22

Sinusoidal ac Circuit Analysis

22.1

a)   Their magnitudes are

$$|z_1| = \sqrt{3^2 + 4^2} = \sqrt{25} = 5 \quad \text{and} \quad |z_2| = \sqrt{2^2 + (-5)^2} = \sqrt{29} \approx 5.4.$$

b)   The polar angles are

$$\theta_1 = \arctan\left(\frac{\text{Im } z_1}{\text{Re } z_1}\right) = \arctan\left(\frac{4}{3}\right) = 0.927 \text{ rad}, \text{ and}$$

$$\theta_2 = \arctan\left(\frac{\text{Im } z_2}{\text{Re } z_2}\right) = \arctan\left(\frac{-5}{2}\right) = -1.19 \text{ rad}.$$

So, using the magnitudes from part a), the polar forms are

$$z_1 = 5 \angle (0.927 \text{ rad}) \quad \text{and} \quad z_2 = \sqrt{29} \angle (-1.19 \text{ rad})$$

c)   With the known magnitudes and polar angles, the exponential forms are

$$z_1 = 5e^{i(0.927 \text{ rad})} \quad \text{and} \quad z_2 = \sqrt{29}e^{-i(1.19 \text{ rad})}$$

d)   The locations of the two complex numbers in the complex plane are shown below.

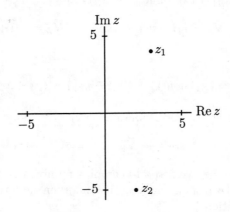

22.5

a)

$$
\begin{aligned}
(1-i)^4 &= [(1-i)^2]^2 \\
&= (1-2i-1)^2 \\
&= (-2i)^2 \\
&= 4i^2 = -4.
\end{aligned}
$$

b)

$$
\begin{aligned}
(\sqrt{2}-i) - i(1-i\sqrt{2}) &= \sqrt{2} - i - i + i^2\sqrt{2} \\
&= \sqrt{2} - i - i - \sqrt{2} = -2i.
\end{aligned}
$$

c)

$$
\begin{aligned}
\frac{5}{(1-i)(2-i)(3-i)} &= \frac{5}{(2-i-i2-1)(3-i)} \\
&= \frac{5}{(1-3i)(3-i)} \\
&= \frac{5}{3-i-9i+3i^2} \\
&= \frac{5}{-10i} \\
&= \left(\frac{5}{-10i}\right)\left(\frac{i}{i}\right) = \frac{i}{2}.
\end{aligned}
$$

22.9 Let the complex number $z = x + iy$ correspond to the vector $x\hat{\mathbf{i}} + y\hat{\mathbf{j}}$. That is, z corresponds to $(\operatorname{Re} z)\hat{\mathbf{i}} + (\operatorname{Im} z)\hat{\mathbf{j}}$. In order to indicate this we write

$$
z \longleftrightarrow (\operatorname{Re} z)\hat{\mathbf{i}} + (\operatorname{Im} z)\hat{\mathbf{j}}
$$

Consider two complex numbers

$$
z_1 = x_1 + iy_1 \qquad \text{and} \qquad z_2 = x_2 + iy_2.
$$

Their sum is

$$
z_1 + z_2 = (x_1 + iy_1) + (x_2 + iy_2) = (x_1 + x_2) + i(y_1 + y_2).
$$

Now consider the two *corresponding* two-dimensional vectors

$$
\vec{\mathbf{V}}_1 = x_1\hat{\mathbf{i}} + y_1\hat{\mathbf{j}} \qquad \text{and} \qquad \vec{\mathbf{V}}_2 = x_2\hat{\mathbf{i}} + y_2\hat{\mathbf{j}}.
$$

Their vector sum is

$$
\vec{\mathbf{V}}_1 + \vec{\mathbf{V}}_2 = \left(x_1\hat{\mathbf{i}} + y_1\hat{\mathbf{j}}\right) + \left(x_2\hat{\mathbf{i}} + y_2\hat{\mathbf{j}}\right) = (x_1 + x_2)\hat{\mathbf{i}} + (y_1 + y_2)\hat{\mathbf{j}}.
$$

Notice that this is precisely the vector corresponding to $z_1 + z_2$. So we've shown that

$$
z_1 \longleftrightarrow \vec{\mathbf{V}}_1 \qquad \text{and} \qquad z_2 \longleftrightarrow \vec{\mathbf{V}}_2 \qquad \Longrightarrow \qquad (z_1 + z_2) \longleftrightarrow (\vec{\mathbf{V}}_1 + \vec{\mathbf{V}}_2).
$$

Therefore, it doesn't matter whether we first add complex numbers and then form the corresponding vector, or whether we first form the corresponding vectors and then add them. In mathematician's lingo, the correspondence "preserves addition."

In exactly the same way, we can show that the correspondence also preserves subtraction:

$$z_1 \longleftrightarrow \vec{V}_1 \quad \text{and} \quad z_2 \longleftrightarrow \vec{V}_2 \quad \Longrightarrow \quad (z_1 - z_2) \longleftrightarrow (\vec{V}_1 - \vec{V}_2).$$

Such a correspondence is called an "isomorphism," and the two mathematical structures, complex numbers and two dimensional vectors, are called "isomorphic" — at least so far as addition and subtraction are concerned. In a sense, if all we cared about were addition and subtraction, then "complex number" and "two dimensional vector" are really just two different names for the same thing. The notion of isomorphism makes this sense precise.

22.13 The inductive reactance is

$$\mathcal{X}_L = \omega L.$$

So, the SI units of the inductive reactance are:

$$(\,\text{rad/s}\,)(\,\text{H}\,) = \frac{\text{H}}{\text{s}}.$$

We have to show that $\dfrac{\text{H}}{\text{s}} = \Omega$. By its definition (see Equation 21.30 on page 971 of the text), one henry is

$$\text{H} = \text{T} \cdot \left(\frac{\text{m}^2}{\text{A}}\right).$$

By its definition (see text, page 898), one tesla is

$$\text{T} = \frac{\text{N} \cdot \text{s}}{\text{C} \cdot \text{m}}$$

Combining these definitions,

$$\frac{\text{H}}{\text{s}} = \frac{\text{T} \cdot \left(\dfrac{\text{m}^2}{\text{A}}\right)}{\text{s}} = \frac{\left(\dfrac{\text{N} \cdot \text{s}}{\text{C} \cdot \text{m}}\right) \cdot \left(\dfrac{\text{m}^2}{\text{A}}\right)}{\text{s}} = \frac{\text{N} \cdot \text{m}}{\text{C} \cdot \text{A}}$$

Now one newton·meter is a joule, $\text{N} \cdot \text{m} = \text{J}$, one joule per coulomb is a volt, $\dfrac{\text{J}}{\text{C}} = \text{V}$, and one volt per ampere is an ohm, $\dfrac{\text{V}}{\text{A}} = \Omega$. Hence,

$$\frac{\text{H}}{\text{s}} = \frac{\text{N} \cdot \text{m}}{\text{C} \cdot \text{A}} = \frac{\text{J}}{\text{C} \cdot \text{A}} = \frac{\text{V}}{\text{A}} = \Omega.$$

22.17 The inductive reactance is

$$\mathcal{X}_L = \omega L = 2\pi\nu L = 2\pi\nu(0.500\ \text{H}) = (3.14\ \Omega\cdot\text{s}\,)\nu.$$

A graph of $\mathcal{X}_L$ versus ν over the interval $0\ \text{Hz} \le \nu \le 1.00\ \text{kHz}$ is linear with zero intercept and slope $3.14\ \Omega\cdot\text{s}$. When $\nu = 1.00\ \text{kHz}$, the inductive reactance is $3.14\ \text{k}\Omega$. Here is the graph.

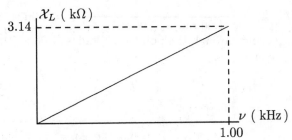

22.21   The capacitive reactance is

$$\mathcal{X}_C = \frac{1}{\omega C} = \frac{1}{2\pi\nu C} \implies \nu = \frac{1}{2\pi \mathcal{X}_c C}$$

Thus, for a 50 μF capacitor, the capacitive reactance will be 10.0 Ω if

$$\nu = \frac{1}{2\pi(10.0\ \Omega)(50 \times 10^{-6}\ \text{F})} = 3.2 \times 10^2\ \text{Hz}.$$

Since the higher the frequency, the lower the capacitive reactance, the capacitive reactance will be *less* than 10 Ω for frequencies *greater* than 3.2×10^2 Hz.

22.25

a)   The equivalent impedance of two circuit elements in series is their sum. Therefore we need $\mathcal{Z}_L + \mathcal{Z}_C = 0\ \Omega$. Let's see what this implies for the frequency ν.

$$\mathcal{Z}_L + \mathcal{Z}_C = 0\ \Omega \implies i\omega L + \frac{1}{i\omega C} = 0\ \Omega \implies (i\omega L)(i\omega C) + 1 = 0 \implies -\omega^2 LC + 1 = 0 \implies \omega = \frac{1}{\sqrt{LC}}.$$

Therefore

$$\nu = \frac{\omega}{2\pi} = \frac{1}{2\pi\sqrt{LC}}.$$

b)   For the inductor and capacitor in parallel, the equivalent impedance is

$$Z_{\text{eq}} = \frac{\mathcal{Z}_L \mathcal{Z}_C}{\mathcal{Z}_L + \mathcal{Z}_C}.$$

This cannot be zero unless either $\mathcal{Z}_L$ or $\mathcal{Z}_C$ is zero. The impedance of a capacitor is never zero. The inductor has zero impedance only if the current is direct current or if it has zero inductance.

Hence the only way that a parallel combination of inductor and capacitor can have zero impedance is if $\nu = 0$ Hz, so the current is direct current, or if the inductor is really just a pure conductor (short circuit).

22.29

a)   and

b)   The capacitive reactance is

$$\mathcal{X}_C = \frac{1}{\omega C}$$

Take logarithms to the base 10 of both sides.

$$\log_{10} \mathcal{X}_C = -\log_{10} C - \log_{10} \omega.$$

Notice that $\log_{10} \mathcal{X}_C$ is a linear function of $\log_{10} \omega$, with a slope of -1. Therefore the graph of $\mathcal{X}_C$ as a function of ω on double log paper will be a straight line with slope of -1. Since its graph is a straight line of known slope, we need to know its value at only one point in order to draw the graph.

When $\omega = 1.00$ rad/s, $\mathcal{X}_C = \dfrac{1}{(1.00\ \text{rad/s})(100 \times 10^{-6}\ \text{F})} = 1.00 \times 10^4\ \Omega$. Hence the graph will be a straight line of slope -1 passing through the point $(1.00\ \text{rad/s}, 1.00 \times 10^4\ \Omega)$.

The inductive reactance is

$$\mathcal{X}_L = \omega L.$$

Take logarithms to the base 10 of both sides.

$$\log_{10} \mathcal{X}_L = \log_{10} L + \log_{10} \omega.$$

22.29.

Therefore $\log_{10} \mathcal{X}_C$ is also linear function of $\log_{10} \omega$. Its slope is $+1$. So, the graph of $\mathcal{X}_L$ as a function of ω on double log paper will be a straight line with slope of $+1$. Since its graph is a straight line of known slope, we need to know its value at only one point in order to draw the graph.

When $\omega = 1.00$ rad/s, $\mathcal{X}_L = (1.00 \text{ rad/s})(0.500 \text{ H}) = 0.500 \, \Omega$. Hence the graph will be a straight line of slope $+1$ passing through the point $(1.00 \text{ rad/s}, 0.500 \, \Omega)$.

Finally, the $1.00 \text{ k}\Omega$ resistor has an impedance of $1.00 \text{ k}\Omega$ independent of frequency, so the graph of its impedance will be a horizontal line at a height of $1.00 \times 10^3 \, \Omega$.

Here are the graphs.

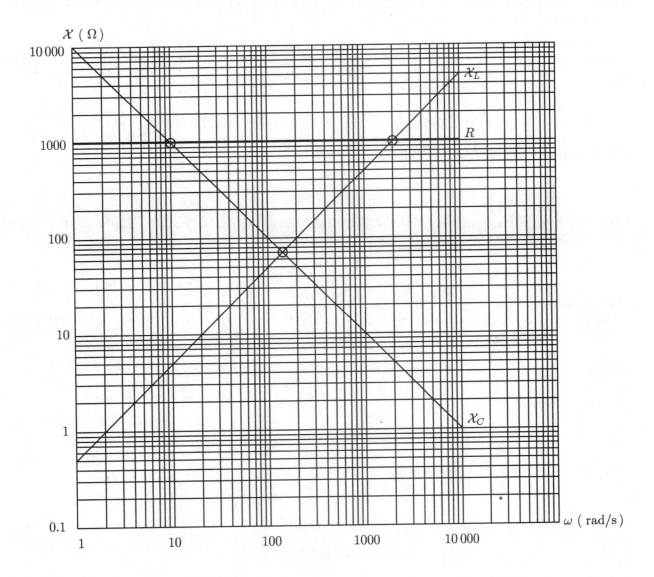

c) This angular frequency can be read from the graph (we've circled it). It can also be obtained by solving for the angular frequency ω as a function of inductive reactance $\mathcal{X}_L$ and substituting $1.00 \text{ k}\Omega$ for $\mathcal{X}_L$:

$$\mathcal{X}_L = \omega L \implies \omega = \frac{\mathcal{X}_L}{L} = \frac{1.00 \times 10^3 \, \Omega}{0.500 \text{ H}} = 2.00 \times 10^3 \text{ rad/s}.$$

The corresponding frequency in Hz is

$$\nu = \frac{\omega}{2\pi} = \frac{2.00 \times 10^3 \text{ rad/s}}{2\pi} = 318 \text{ Hz}.$$

d) This angular frequency can also be read from the graph (we've circled it). It can also be obtained by solving for the angular frequency ω as a function of capacitive reactance $\mathcal{X}_C$ and substituting 1.00 kΩ for $\mathcal{X}_C$:

$$\mathcal{X}_C = \frac{1}{\omega C} \implies \omega = \frac{1}{\mathcal{X}_C C} = \frac{1}{(1.00 \times 10^3\ \Omega)(100 \times 10^{-6}\ F)} = 10\ \text{rad/s}.$$

The corresponding frequency in Hz is

$$\nu = \frac{\omega}{2\pi} = \frac{10\ \text{rad/s}}{2\pi} = 1.59\ \text{Hz}.$$

e) This angular frequency can also be read from the graph; we've circled where the two graphs cross. It can also be obtained by setting the two reactances equal to each other and solving for the angular frequency ω.

$$\mathcal{X}_C = \frac{1}{\omega C} \implies \omega L = \frac{1}{\omega C} \implies \omega = \sqrt{\frac{1}{LC}} = \sqrt{\frac{1}{(0.500\ H)(100 \times 10^{-6}\ F)}} = 141\ \text{rad/s}.$$

The corresponding frequency in Hz is

$$\nu = \frac{\omega}{2\pi} = \frac{141\ \text{rad/s}}{2\pi} = 22.4\ \text{Hz}.$$

22.33 Let V_0 be the peak value of the ac potential difference across the resistor. The ac power absorbed by the resistor is

$$\frac{1}{2}V_0 I_0 = \frac{1}{2}(I_0 R)I_0 = \frac{1}{2}I_0^2 R.$$

Let I_{dc} be the dc current that results in the same power absorbed by the resistor. Then

$$I_{dc}^2 R = \frac{1}{2}I_0^2 R \implies I_{dc} = \frac{I_0}{\sqrt{2}}$$

22.37

a) The angular frequency of the source is the coefficient of t in the argument of the cosine, so

$$\omega = 377\ \text{rad/s} \implies \nu = \frac{\omega}{2\pi} = \frac{377\ \text{rad/s}}{2\pi\ \text{rad}} = 60.0\ \text{Hz}.$$

b) The impedance of an inductor is

$$\mathcal{Z}_L = i\omega L = i(377\ \text{rad/s})(0.150\ H) = i(56.6\ \Omega).$$

c) The independent voltage source phaser is

$$\mathbf{V}(t) = (170\ \text{V}) \angle (377\ \text{rad/s})t = (170\ \text{V})e^{i(377\ \text{rad/s})t}$$

d) Choose the direction of the current phaser as shown below. The current enters the positive polarity terminal of the inductor.

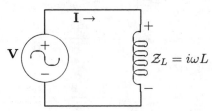

22.37.

e) Apply the KVL clockwise around the loop.

$$-\mathbf{V} + \mathbf{I}\mathcal{Z}_L = 0 \text{ V} \implies$$

$$\mathbf{I} = \frac{\mathbf{V}}{\mathcal{Z}_L} = \frac{\mathbf{V}}{i(56.6 \ \Omega)} = \frac{-i\mathbf{V}}{56.6 \ \Omega} = \frac{e^{-i(\pi/2 \text{ rad})}(170 \text{ V})e^{i(377 \text{ rad/s})t}}{56.6 \ \Omega}$$

$$= (3.00 \text{ A})e^{i(-\pi/2 \text{ rad} + (377 \text{ rad/s})t)} = (3.00 \text{ A}) \angle \left(-\frac{\pi}{2} \text{ rad} + (377 \text{ rad/s})t\right)$$

f) The potential difference phaser across the inductor is

$$\mathbf{V}_L = \mathbf{I}\mathcal{Z}_L = (3.00 \text{ A})e^{i(-\pi/2 \text{ rad} + (377 \text{ rad/s})t)}(i56.6 \ \Omega) = (3.00 \text{ A})e^{i(-\pi/2 \text{ rad} + (377 \text{ rad/s})t)}e^{i\pi/2 \text{ rad}}(56.6 \ \Omega)$$

$$= (170 \text{ V})e^{i(377 \text{ rad/s})t} = (170 \text{ V}) \angle [(377 \text{ rad/s})t].$$

g) The real current through the inductor is the real part of the current phaser.

$$I(t) = (3.00 \text{ A})\cos\left(-\frac{\pi}{2} \text{ rad} + (377 \text{ rad/s})t\right).$$

h) The real potential difference across the inductor is the real part of the potential difference phaser across it.

$$V_L(t) = (170 \text{ V})\cos[(377 \text{ rad/s})t].$$

i) The phase difference between the potential difference across the inductor and the current through it is

$$\beta = \text{phase difference} = (377 \text{ rad/s})t - \left(-\frac{\pi}{2} \text{ rad} + (377 \text{ rad/s})t\right) = \frac{\pi}{2} \text{ rad}.$$

j) The peak current through the inductor is the coefficient of the cosine term in part g).

$$I_{\text{peak}} = 3.00 \text{ A}.$$

k) The rms current through the inductor is the peak value divided by $\sqrt{2}$.

$$I_{\text{rms}} = \frac{3.00 \text{ A}}{\sqrt{2}} = 2.12 \text{ A}.$$

l) The peak potential difference across the inductor is the coefficient of the cosine term in part h).

$$V_{L \text{ peak}} = 170 \text{ V}.$$

m) The rms potential difference across the inductor is the peak value divided by $\sqrt{2}$.

$$V_{L \text{ rms}} = \frac{170 \text{ V}}{\sqrt{2}} = 120 \text{ V}.$$

n) The power factor associated with the inductor is the cosine of the phase angle difference β between the potential difference across the inductor and the current through it.

$$\text{power factor} = \cos\beta = \cos\left(\frac{\pi}{2} \text{ rad}\right) = 0.$$

o) The average power absorbed by the inductor is

$$\langle P \rangle = V_{L \text{ rms}}I_{\text{rms}}\cos\beta = V_{L \text{ rms}}I_{\text{rms}}0 = 0 \text{ W}.$$

22.41   Choose the current direction to be out of the positive terminal of the voltage source, and mark the polarities of the impedances accordingly. Apply the KVL clockwise around the loop:

$$-\mathbf{V} + \mathbf{I}R + \mathbf{I}\left(\frac{-i}{\omega C}\right) = 0 \text{ V} \implies$$

$$\mathbf{I} = \frac{\mathbf{V}}{R - \dfrac{i}{\omega C}} = \frac{\mathbf{V}\omega C}{\omega RC - i} = \frac{V_0 \omega C \ \angle \ \omega t}{\sqrt{(\omega RC)^2 + (-1)^2} \ \angle \ \arctan\left(\dfrac{-1}{\omega RC}\right)}$$

$$= \frac{V_0 \omega C}{\sqrt{(\omega RC)^2 + 1}} \ \angle \ \left(\omega t + \arctan\left(\frac{1}{\omega RC}\right)\right)$$

The potential difference phaser across the capacitor is

$$\mathbf{V}_C = \mathbf{I}\mathcal{Z}_C = \left[\frac{V_0 \omega C}{\sqrt{(\omega RC)^2 + 1}} \ \angle \ \left(\omega t + \arctan\left(\frac{1}{\omega RC}\right)\right)\right]\left(-\frac{i}{\omega C}\right)$$

$$= \left[\frac{V_0}{\sqrt{(\omega RC)^2 + 1}} \ \angle \ \left(\omega t + \arctan\left(\frac{1}{\omega RC}\right)\right)\right](-i)$$

$$= \left[\frac{V_0}{\sqrt{(\omega RC)^2 + 1}} \ \angle \ \left(\omega t + \arctan\left(\frac{1}{\omega RC}\right)\right)\right]\left(\angle \ -\frac{\pi}{2} \text{ rad}\right)$$

$$= \frac{V_0}{\sqrt{(\omega RC)^2 + 1}} \ \angle \ \left(\omega t + \arctan\left(\frac{1}{\omega RC}\right) - \frac{\pi}{2} \text{ rad}\right).$$

This has a peak value of

$$V_{C\text{ peak}} = \frac{V_0}{\sqrt{(\omega RC)^2 + 1}}.$$

We want this to be half the peak value of the source:

$$\frac{V_0}{\sqrt{(\omega RC)^2 + 1}} = \frac{V_0}{2} \implies \sqrt{(\omega RC)^2 + 1} = 2 \implies (\omega RC)^2 + 1 = 4 \implies \omega = \frac{\sqrt{3}}{RC}.$$

The voltage gain is

$$\text{voltage gain} = \frac{V_{C\text{ peak}}}{V_0} = \frac{\dfrac{V_0}{2}}{V_0} = \frac{1}{2}.$$

22.45

a)   From Equation 22.74 on page 1025 of the text, for the voltage gain of the low pass filter is

$$\text{voltage gain} = \frac{1}{\sqrt{1 + (\omega RC)^2}} = \frac{1}{\sqrt{1 + (2\pi\nu RC)^2}}$$

Since

$$(2\pi\nu RC)^2 = [2\pi(1.00 \times 10^3 \ \Omega)(0.0500 \times 10^{-6} \text{ F})]^2\nu^2 = (9.86 \times 10^{-8} \text{ s}^2)\nu^2,$$

the voltage gain is

$$\text{voltage gain} = \frac{1}{\sqrt{1 + (9.86 \times 10^{-8} \text{ s}^2)\nu^2}}.$$

A Bode plot is shown below:

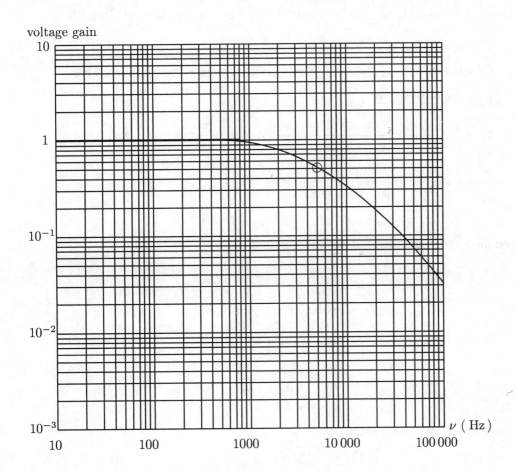

voltage gain

b)   From the graph, when the voltage gain is 0.50, the frequency is

$$\nu \approx 5 \times 10^3 \text{ Hz}.$$

To *calculate* the frequency, set the expression for the voltage gain equal to 0.50 and solve for ν.

$$0.50 = \frac{1}{\sqrt{1 + (9.86 \times 10^{-8} \text{ s}^2)\nu^2}} \implies 2.0 = \sqrt{1 + (9.86 \times 10^{-8} \text{ s}^2)\nu^2}$$

$$\implies 4.0 = 1 + (9.86 \times 10^{-8} \text{ s}^2)\nu^2 \implies \nu = 5.5 \times 10^3 \text{ Hz}.$$

22.49

a)   Here's the circuit in the complex domain.

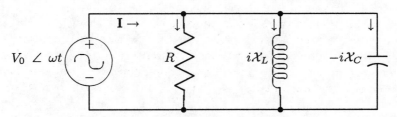

b)   Find the current phaser through each circuit element from the complex form of Ohm's law:

$$\mathbf{V} = \mathbf{I}\mathcal{Z} \implies \mathbf{I} = \frac{\mathbf{V}}{\mathcal{Z}}.$$

Here, $\mathbf{V} = V_0 \angle \omega t$ for all of the circuit elements, since they are connected in parallel. Thus

$$\mathbf{I}_R = \frac{V_0 \angle \omega t}{R} = \frac{V_0}{R} \angle \omega t,$$

$$\mathbf{I}_L = \frac{V_0 \angle \omega t}{i\mathcal{X}_L} = \frac{V_0 \angle \omega t}{\mathcal{X}_L \angle \frac{\pi}{2} \text{ rad}} = \frac{V_0}{\mathcal{X}_L} \angle \left(\omega t - \frac{\pi}{2}\text{ rad}\right), \qquad \text{and}$$

$$\mathbf{I}_C = \frac{V_0 \angle \omega t}{-i\mathcal{X}_C} = \frac{V_0 \angle \omega t}{\mathcal{X}_C \angle -\frac{\pi}{2}\text{ rad}} = \frac{V_0}{\mathcal{X}_C} \angle \left(\omega t + \frac{\pi}{2}\text{ rad}\right).$$

c)   Since the impedances are in parallel, the equivalent impedance satisfies

$$\frac{1}{\mathcal{Z}_{\text{eq}}} = \frac{1}{\mathcal{Z}_R} + \frac{1}{\mathcal{Z}_L} + \frac{1}{\mathcal{Z}_C}$$

$$= \frac{1}{R} + \frac{1}{i\mathcal{X}_L} + \frac{1}{-i\mathcal{X}_C}$$

$$= \frac{1}{R} + \frac{-i}{\mathcal{X}_L} + \frac{i}{\mathcal{X}_C}$$

$$= \frac{1}{R} + i\left(\frac{1}{\mathcal{X}_C} - \frac{1}{\mathcal{X}_L}\right).$$

d)   The current phaser from the source satisfies

(1) $$\mathbf{V} = \mathbf{I}\mathcal{Z}_{\text{eq}} \implies \mathbf{I} = \mathbf{V}\frac{1}{\mathcal{Z}_{\text{eq}}}.$$

We may rewrite the answer to part c) in polar form in terms of an angle θ.

$$\frac{1}{\mathcal{Z}_{\text{eq}}} = \frac{1}{R} + i\left(\frac{1}{\mathcal{X}_C} - \frac{1}{\mathcal{X}_L}\right) = \sqrt{\left(\frac{1}{R}\right)^2 + \left(\frac{1}{\mathcal{X}_C} - \frac{1}{\mathcal{X}_L}\right)^2} \angle \theta.$$

The actual value of θ depends upon the ratio of the imaginary part $\left(\dfrac{1}{\mathcal{X}_C} - \dfrac{1}{\mathcal{X}_L}\right)$ to the real part $\dfrac{1}{R}$, but we won't need to know what the actual value of θ is. (Whew!) So (1) becomes

$$\mathbf{I} = \mathbf{V}\frac{1}{\mathcal{Z}_{\text{eq}}} = (V_0 \angle \omega t)\left(\sqrt{\left(\frac{1}{R}\right)^2 + \left(\frac{1}{\mathcal{X}_C} - \frac{1}{\mathcal{X}_L}\right)^2} \angle \theta\right) = V_0\sqrt{\left(\frac{1}{R}\right)^2 + \left(\frac{1}{\mathcal{X}_C} - \frac{1}{\mathcal{X}_L}\right)^2} \angle (\theta + \omega t)$$

The actual current is the real part of this phaser. The times t at which it will be a maximum are when $\cos(\omega t + \theta) = 1$, that is, when $\omega t + \theta$ is a multiple of 2π. At those times the current will be

$$I_{\text{peak}} = V_0\sqrt{\left(\frac{1}{R}\right)^2 + \left(\frac{1}{\mathcal{X}_C} - \frac{1}{\mathcal{X}_L}\right)^2}.$$

e) Actually, *there is no maximum current for this circuit.*

One can see this qualitatively by looking at the circuit. The higher the frequency of the source, the lower the reactance of the capacitor, so one can make the current through the capacitor arbitrarily high by increasing the frequency of the source. Of course at very high frequencies, the inductor acts more and more like an open circuit — but this doesn't matter. At high frequencies, the current will just zip through the capacitor.

On the other hand, at very low frequencies, the inductor acts more and more like a short circuit. Although the capacitor acts as an open circuit for dc, that doesn't matter to the current. The current can just zip through the inductor.

Mathematically, the expression

$$I_{\text{peak}} = V_0 \sqrt{\left(\frac{1}{R}\right)^2 + \left(\frac{1}{X_C} - \frac{1}{X_L}\right)^2}.$$

for the peak current, has no maximum. It can be made arbitrarily large by making either X_C small (high frequencies), or by making X_L small (low frequencies).

The expression does have a *minimum*, however. The current is *minimized* if $X_L = X_C$. That's because the squared term $\left(\frac{1}{X_C} - \frac{1}{X_L}\right)^2$ is always nonnegative, but is equal to zero only when $X_L = X_C$. To find the corresponding angular frequency, set $X_L = X_C$ and turn the crank:

$$X_L = X_C \implies \omega L = \frac{1}{\omega C} \implies \omega = \frac{1}{\sqrt{LC}}.$$

It is interesting, that the angular frequency which *maximizes* the current in a *series RLC* circuit, is the same angular frequency that *minimizes* it in a *parallel RLC* circuit.

Chapter 23

Geometric Optics

23.1 The light intensity at a distance r form the Sun is $I = \dfrac{L}{4\pi r^2}$, where L is the luminosity of the Sun. So, for the Earth and for Saturn

$$I_{\text{Earth}} = \frac{L}{4\pi r_{\text{Earth}}^2} \quad \text{and} \quad I_{\text{Saturn}} = \frac{L}{4\pi r_{\text{Saturn}}^2}.$$

Divide the second equation by the first:

$$\frac{I_{\text{Saturn}}}{I_{\text{Earth}}} = \frac{\dfrac{L}{4\pi r_{\text{Saturn}}^2}}{\dfrac{L}{4\pi r_{\text{Earth}}^2}} = \frac{r_{\text{Earth}}^2}{r_{\text{Saturn}}^2} \implies I_{\text{Saturn}} = I_{\text{Earth}} \frac{r_{\text{Earth}}^2}{r_{\text{Saturn}}^2} = (1.36 \text{ kW/m}^2) \frac{r_{\text{Earth}}^2}{(9.54 r_{\text{Earth}})^2} = 14.9 \text{ W/m}^2.$$

23.5 We'll first find an expression for the *density* of the solar energy, and then multiply the density by the volume of one cubic kilometer of space.

Consider a cube in space located a distance r from the Sun, and let $\Delta\ell$ be the length of any one of its edges. We assume that r is very big relative to $\Delta\ell$, so we don't care exactly which part of the cube is at a distance r from the Sun. Assume that the cube is nicely oriented with respect to the Sun, so that one of its faces (call it the front face) is perpendicular to the rays coming from the Sun. Then $I(\Delta\ell)^2$ is the *power* of the Sun on the front face of the cube, where I is the intensity of the solar light on the cube. This is the rate at which energy is passing through the cube. Since the light must travel a distance ℓ before leaving the cube, each unit of energy is in the cube for a length of time $\dfrac{\Delta\ell}{c}$, where c is the speed of light. Then,

$$\text{total energy on the way through the cube} = I(\Delta\ell^2)\frac{\Delta\ell}{c} = \frac{I}{c}(\Delta\ell)^3.$$

An analogy may help here: Think of a college where each entering student graduates four years after entrance. Then at any time, the total number of students in the college is equal to four times the number of students entering each year. Thus, if each entering class has 600 students, then there will be $600 \times 4 = 2400$ students in the college. In the case of the cube, "students" arrive at the rate of $I(\Delta\ell)^2$, and they stay in school for a length of time $\dfrac{\Delta\ell}{c}$, so the total number in college is $I(\Delta\ell^2)\dfrac{\Delta\ell}{c} = \dfrac{I}{c}(\Delta\ell)^3$.

The density of the solar energy is the amount of energy in the cube divided by the volume of the cube.

$$\rho_{\text{solar energy}} = \frac{\left(\dfrac{I}{c}(\Delta\ell)^3\right)}{(\Delta\ell)^3} = \frac{I}{c}.$$

The light intensity at distance r from the Sun is

$$I = \frac{L}{4\pi r^2},$$

217

where L is the luminosity of the Sun. Therefore

$$\rho_{\text{solar energy}} = \frac{L}{4\pi r^2 c}.$$

Now let V_{chunk} be the volume of any chunk of space, located at a distance r from the Sun, where the chunk is small enough or symmetrical enough so that r doesn't vary appreciably within the chunk. Then the amount of solar energy in the chunk is

$$\text{solar energy in chunk} = \frac{L}{4\pi r^2 c} V_{\text{chunk}}.$$

Notice that it doesn't particularly matter how the chunk of space is shaped — in particular it doesn't have to be a cube facing into the Sun.

Now plug in some numbers. Suppose the chunk is one cubic kilometer. Then $V_{\text{chunk}} = 10^9 \text{ m}^3$, so

$$\text{solar energy in chunk} = \frac{L}{4\pi r^2 c}(10^9 \text{ m}^3) = \frac{3.83 \times 10^{26} \text{ W}}{4\pi (1.496 \times 10^{11} \text{ m})^2 (3.00 \times 10^8 \text{ m/s})}(10^9 \text{ m}^3) = 4.54 \times 10^3 \text{ J}.$$

23.9   Here's a slightly modified picture.

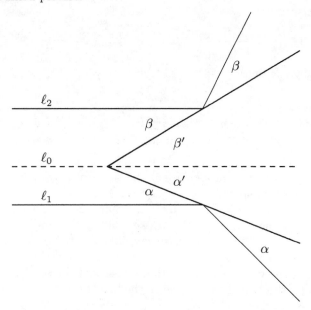

The lines ℓ_1 and ℓ_2 are the two parallel incident light rays. We've added the dashed line ℓ_0 to the picture. It is a line parallel to both the incident light rays and passes through the point where the mirrors are joined. From the law of reflection, the two angles marked α are equal. (The angle α is the complement of the angle of incidence of this ray.) Similarly for the two angles marked β. We haven't included the angle ϕ in this picture, but it is the same as $\alpha' + \beta'$:

$$\phi = \alpha' + \beta'.$$

The angle between the two reflected rays is $\alpha + \alpha' + \beta' + \beta$. Since we want to show that this is 2ϕ, we need to show that

$$\alpha + \alpha' + \beta' + \beta = 2(\alpha' + \beta').$$

Since ℓ_1 and ℓ_0 are parallel, $\alpha = \alpha'$, and since ℓ_2 and ℓ_0 are parallel, $\beta = \beta'$. Therefore,

$$\alpha + \alpha' + \beta' + \beta = \alpha' + \alpha' + \beta' + \beta' = 2(\alpha' + \beta'),$$

as was to be shown.

23.13

a) With the eye at P in the diagram (Figure P.13 on page 1087 of the text), the ray making the smallest angle with the axis is θ_1. Since this is the smallest angle, the rays that they form will appear to be the inside circle. By symmetry, this ray is reflected from the center of the length of the tube.

b) Again by symmetry, the ray making two reflections from the side of the tube reaches the eye at P making angle θ_2 with the axis if the reflections are at distances $\dfrac{L}{4}$ and $\dfrac{3L}{4}$ from the hole in the left end of the tube.

c) The ray making angle θ_3 with the axis at P, is reflected three times from the sides of the tube; from the symmetry, the reflections are at points $\dfrac{L}{6}$, $\dfrac{3L}{6}$, and $\dfrac{5L}{6}$ from the hole at the left end.

d) From the geometry, the ring closest to the axis [the ring from part a)] reaches the eye at an angle with tangent

$$\tan\theta_1 = \frac{\left(\dfrac{D}{2}\right)}{\left(\dfrac{L}{2}\right)} = \frac{D}{L}.$$

The rays forming the second ring reach the eye making an angle with the axis with tangent

$$\tan\theta_2 = \frac{\left(\dfrac{D}{2}\right)}{\left(\dfrac{L}{4}\right)} = \frac{2D}{L}.$$

The rays forming the third ring reach the eye making an angle with the axis with tangent

$$\tan\theta_3 = \frac{\left(\dfrac{D}{2}\right)}{\left(\dfrac{L}{6}\right)} = \frac{3D}{L}.$$

In general, the rays forming the mth ring reach the eye making an angle with the axis with tangent

$$\tan\theta_m = \frac{\left(\dfrac{D}{2}\right)}{\left(\dfrac{L}{2m}\right)} = \frac{mD}{L}.$$

23.17 Use the geometry shown below (based on Figure P.17 on page 1088 of the text). We need to find x.

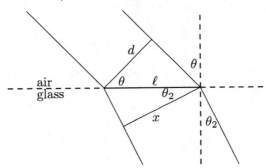

We see that $x = \ell\cos\theta_2$, so our approach is to express ℓ and $\cos\theta_2$ in terms of the known quantities d, n, and θ.

From the top triangle $\dfrac{d}{\ell} = \cos\theta$, so $\ell = \dfrac{d}{\cos\theta}$, and therefore

(1)
$$x = \frac{d\cos\theta_2}{\cos\theta}.$$

Now for $\cos\theta_2$. We use the law of refraction, $n_1\sin\theta_1 = n_2\sin\theta_2$. In this application if we let $\theta_1 = \theta$, then $n_1 = 1.00$, for air, and $n_2 = n$. Thus $\sin\theta = n\sin\theta_2$ so

$$\sin\theta_2 = \frac{\sin\theta}{n}.$$

Now use the identity $\sin^2\theta_2 + \cos^2\theta_2 = 1$ to find

$$\cos\theta_2 = \pm\sqrt{1 - \sin^2\theta_2} = \pm\sqrt{1 - \left(\frac{\sin\theta}{n}\right)^2}.$$

Choose the positive root, since $0 \le \theta_2 \le 90°$, and substitute into (1),

$$x = \frac{d\sqrt{1 - \left(\dfrac{\sin\theta}{n}\right)^2}}{\cos\theta}.$$

This can be rewritten in a number of simpler ways, one of which is to multiply top and bottom by n and put everything inside the square root:

$$x = \frac{d\sqrt{1 - \left(\dfrac{\sin\theta}{n}\right)^2}}{\cos\theta} = d\sqrt{\frac{n^2 - \sin^2\theta}{n^2\cos^2\theta}}.$$

23.21 In order to apply the law of refraction, the angles need to be measured from the *normal* to the surface. However, the given angles, $30°$ and ϕ, are *not* measured from the normal. Here's the same picture, but the normal has been added along with a few more angles.

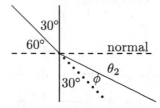

Now apply the law of refraction with $\theta_1 = 60°$, $n_1 = 1.00$ (for air), and $n_2 = 1.50$.

$$n_1\sin\theta_1 = n_2\sin\theta_2 \implies \sin\theta_2 = \frac{n_1}{n_2}\sin\theta_1 = \frac{1.00}{1.50}\sin 60° \implies \theta_2 = 35°.$$

Therefore, from the picture,

$$\phi = 90° - 30° - \theta_2 = 90° - 30° - 35° = 25°.$$

23.25 Here's the picture from the text, with a few additions. We've shown the ray inside the prism parallel to the base, as the problem assumes.

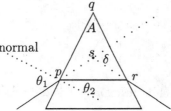

We want to express the angles θ_1 and θ_2 in terms of A and δ, and then apply the law of refraction.

Notice first that δ is an exterior angle of the triangle *psr*, it is therefore equal to the sum of the two remote interior angles $\angle spr$ and $\angle srp$.

$$\delta = \angle spr + \angle srp.$$

Notice also that $\theta_1 = \theta_2 + \angle spr$, so $\angle spr = \theta_1 - \theta_2$. Similarly, $\angle srp = \theta_1 - \theta_2$. Therefore, $\delta = 2(\theta_1 - \theta_2)$, so

(1)
$$\theta_1 = \theta_2 + \frac{1}{2}\delta.$$

Now look at the triangle *pqr*. The base angle $\angle qpr$ is $90° - \theta_2$. Similarly, the other base angle is $90° - \theta_2$. Since the sum of the angles is $180°$, we have $A + (90° - \theta_2) + (90° - \theta_2) = 180°$, so

(2)
$$A = 2\theta_2.$$

It follows from equations (1) and (2) that

$$\theta_2 = \frac{1}{2}A \qquad \text{and} \qquad \theta_1 = \frac{1}{2}(A + \delta)$$

Now apply the law of refraction:

$$n_1 \sin\theta_1 = n_2 \sin\theta_2 \implies 1\sin\left(\frac{1}{2}(A+\delta)\right) = n\sin\left(\frac{1}{2}A\right) \implies n = \frac{\sin\left(\frac{1}{2}(A+\delta)\right)}{\sin\left(\frac{1}{2}A\right)}.$$

23.29

a) Here, with a few additions, is the picture from the text.

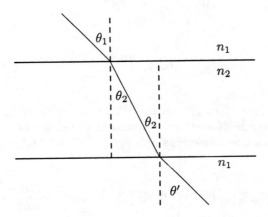

Since the two surfaces are parallel and opposite interior angles are equal, the angle of incidence at the second surface is also θ_2, as we've shown in the picture.

Apply the law of refraction to each surface in succession:

$$n_1 \sin\theta_1 = n_2 \sin\theta_2 \qquad \text{and then} \qquad n_2 \sin\theta_2 = n_1 \sin\theta'.$$

Hence $n_1 \sin\theta_1 = n_1 \sin\theta'$. Since both angles are between $0°$ and $90°$, this means $\theta' = \theta_1$, so the first and last rays are parallel.

b)   In order to find the lateral shift, ℓ, modify the diagram slightly.

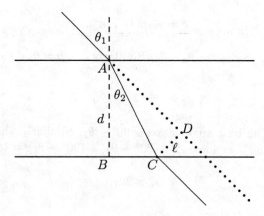

Our strategy is this. We'll first use the geometry to write ℓ entirely in terms of d and the sines and cosines of θ_1 and θ_2. We'll then use the law of refraction and the identity $\cos\theta_2 = \sqrt{1 - \sin^2\theta_2}$ to eliminate sines and cosines of θ_2.

From the right triangle ADC, we have $\ell = \overline{AC}\sin\angle CAD$. Now notice that the angle $\angle BAD$ is equal to θ_1 (since they both are vertical angles), and therefore, $\angle CAD = \theta_1 - \theta_2$. Thus

$$\ell = \overline{AC}\sin(\theta_1 - \theta_2).$$

From the right triangle ABC, $\cos\theta_2 = \dfrac{d}{\overline{AC}}$, so $\overline{AC} = \dfrac{d}{\cos\theta_2}$. Thus,

$$\ell = \frac{d}{\cos\theta_2}\sin(\theta_1 - \theta_2)$$

This expresses ℓ entirely in terms of d, θ_1 and θ_2. However, in order to use the law of refraction, we need to write $\sin(\theta_1 - \theta_2)$ in terms of sines and cosines of the individual angles θ_1 and θ_2. So, use the difference-of-angles formula from trigonometry, $\sin(\theta_1 - \theta_2) = \sin\theta_1\cos\theta_2 - \cos\theta_1\sin\theta_2$. Then,

(1) $$\ell = \frac{d}{\cos\theta_2}(\sin\theta_1\cos\theta_2 - \cos\theta_1\sin\theta_2) = d\sin\theta_1 - \frac{d}{\cos\theta_2}\cos\theta_1\sin\theta_2.$$

Now to eliminate the sines and cosines of θ_2. From the law of refraction, $n_1\sin\theta_1 = n_2\sin\theta_2$, and from the identity $\cos^2\theta_2 = 1 - \sin^2\theta_2$, we have

$$\sin\theta_2 = \frac{n_1}{n_2}\sin\theta_1 \qquad \text{and} \qquad \cos\theta_2 = \sqrt{1 - \left(\frac{n_1}{n_2}\sin\theta_1\right)^2}.$$

Substitute these expressions for the sine and cosine of θ_2 into equation (1) and do some algebra.

$$\ell = d\sin\theta_1 - \frac{d}{\cos\theta_2}\cos\theta_1\sin\theta_2$$

$$= d\sin\theta_1 - \frac{d}{\sqrt{1-\left(\dfrac{n_1}{n_2}\sin\theta_1\right)^2}}\cos\theta_1\frac{n_1}{n_2}\sin\theta_1$$

$$= d\sin\theta_1\left(1 - \frac{\dfrac{n_1}{n_2}\cos\theta_1}{\sqrt{1-\left(\dfrac{n_1}{n_2}\sin\theta_1\right)^2}}\right)$$

$$= d\sin\theta_1\left(1 - \frac{n_1\cos\theta_1}{\sqrt{n_2^2 - n_1^2\sin^2\theta_1}}\right)$$

This is the expression for ℓ that we were supposed to derive.

When $\theta_1 = 0°$, the incident ray is along the normal line and so will be the same as the refracted ray (the angle of refraction then is $0°$), and there will be no lateral shift of the ray as it moves between media with different indices of refraction.

When $n_1 = n_2$, the two media are the same and there is no interface, and therefore no refraction, and no lateral shift of the ray.

23.33

a)   Use the mirror equation:

$$\frac{1}{s} + \frac{1}{s'} = \frac{2}{R} \implies s' = \frac{Rs}{2s - R} \implies s' = \frac{(10.0\text{ cm})(-50\text{ cm})}{2(-50\text{ cm}) - 10.0\text{ cm}} = 4.5\text{ cm}.$$

b)   Use the mirror magnification equation:

$$m = -\frac{s'}{s} = -\frac{4.5\text{ cm}}{-50\text{ cm}} = 0.090.$$

c)   The image distance is positive, so the image is behind the mirror. Therefore, the image is virtual.

d)   The magnification is positive, so the image is right-side-up.

23.37

a)   Since the dentist sees the tooth in back of the mirror, the image is virtual.

b)   Use the mirror magnification equation:

$$m = -\frac{s'}{s}.$$

The image is the same size as the object, so

$$|m| = 1$$

and the image and object distances must have the same absolute value. The image of the tooth in the mirror is virtual, so the image distance s' must be positive. The mirror must either be flat or convex to form such a virtual image, so the radius of curvature is positive. Use this information in the mirror equation.

$$\frac{2}{R} = \frac{1}{s} + \frac{1}{s'} = \frac{1}{s} + \frac{1}{-s} = 0\text{ m}^{-1} \implies R = \infty\text{ m}.$$

The mirror is flat.

23.41   Here's the picture.

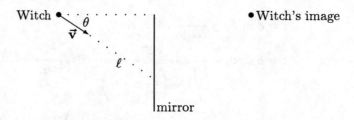

For a plane mirror, the image is located the same distance in back of the mirror as the object is in front of it. The distance between the witch and the mirror is $\ell \cos\theta$, so $2\ell \cos\theta$ is the distance between the witch and her image. The speed at which the image approaches the witch is the time derivative of this distance:

$$\frac{d}{dt} 2\ell \cos\theta = (2\cos\theta)\frac{d\ell}{dt} = (2\cos\theta)v = 2v\cos\theta.$$

23.45   We will find it handy to have the mirror equation $\dfrac{1}{s} + \dfrac{1}{s'} = \dfrac{2}{R}$ solved for s'. After a little algebra, the solution is

(1)
$$s' = \frac{sR}{2s - R}.$$

Set up the standard geometry in order to analyze mirror #1.

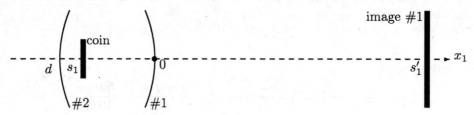

We have one coordinate direction, x_1. Mirror #1 is at the origin, and mirror #2 is at the coordinate position $x_1 = d$. The coin is at coordinate s_1 — later on we'll move it back to d. Notice that because of our coordinate conventions, R, d, and s_1 are all negative. We've drawn the image of the coin in mirror #1 as a virtual image. If s_1 is actually to the left of the focus, then the image should be drawn to the left of mirror #1 as a real image. Apply equation (1) with $s = s_1$, in order to find the position s'_1 of the image.

(2)
$$s'_1 = \frac{s_1 R}{2s_1 - R}.$$

Now the image of mirror #1 becomes the object for mirror #2. Redraw the geometry shown above, but with mirrors interchanged to put the problem into the standard geometry for mirror #2.

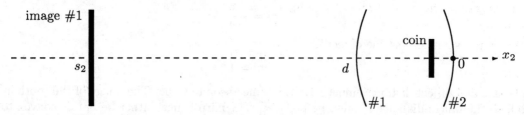

The two coordinate systems are related by $x_2 = d - x_1$. So, in the new coordinate system, the position of image #1 is given by

(3)
$$s_2 = d - s'_1.$$

Use equation (1) with $s = s_2$ to find the position in the x_2 coordinate system of image #2 (the image in mirror #2 of image #1).

$$s_2' = \frac{s_2 R}{2s_2 - R}.$$

Use equation (3) to substitute $d - s_1'$ for s_2. The result is

(4)
$$s_2' = \frac{(d - s_1')R}{2(d - s_1') - R}.$$

Finally, the overall magnification of the coin is the product $m = m_1 m_2$ of the individual magnifications $m_1 = -\frac{s_1'}{s_1}$ and $m_2 = -\frac{s_2'}{s_2}$. Using equation (3) to substitute for s_2, this product is

(5)
$$m = \frac{s_2'}{s_1}\left(\frac{s_1'}{d - s_1'}\right).$$

We're now ready to look at the two different cases, $s_1 = d = \frac{R}{2}$ and $s_1 = d = \frac{3R}{2}$.

Case $s_1 = d = \dfrac{R}{2}$:

Substitute $\frac{R}{2}$ for s_1 in equation (2) and find that $s_1' = \infty$ m. Now take the limit in equation (4) as $s_1' \to \infty$ m:

$$s_2' = \lim_{s_1' \to \infty \text{ m}} \frac{(d - s_1')R}{2(d - s_1') - R} = \lim_{s_1' \to \infty \text{ m}} \frac{R}{2 - \dfrac{R}{d - s_1'}} = \frac{R}{2}.$$

Since $d = \frac{R}{2}$ is the vertex of mirror #1 in the second coordinate system, we've shown that the image of the coin located at the vertex of one mirror (mirror #2) shows up at the vertex of the other mirror (mirror #1).

What about magnification? To find m, take the limit of equation (5) as $s_1' \to \infty$ m.

$$m = \lim_{s_1' \to \infty \text{ m}} \frac{s_2'}{s_1}\left(\frac{s_1'}{d - s_1'}\right) = \left(\frac{s_2'}{s_1}\right)\left(\lim_{s_1' \to \infty \text{ m}} \frac{1}{\dfrac{d}{s_1'} - 1}\right) = \frac{\left(\dfrac{R}{2}\right)}{\left(\dfrac{R}{2}\right)}(-1) = -1.$$

Thus, the image is the same size as the original coin but it is inverted. This does *not* mean that if the coin at mirror # 2 is exposing "heads" to mirror #1, then its image at mirror #1 is "tails." What it *does* mean is that if the coin at mirror # 2 is exposing "heads" to mirror #1, then its image at mirror #1 is also "heads," but it is upside down from the original coin. In the process of going out to infinity and back, the image got turned upside down.

In terms of light rays, here's what happens. Diverging rays from a given spot on the coin travel to mirror #1. Since the coin is at the focus of mirror #1, these rays are reflected back towards mirror #2 as parallel rays. Since these rays are parallel, they are all reflected back by mirror #2 to one spot at the focal plane of mirror #2, where they form a *real* image.

Case $s_1 = d = \dfrac{3R}{2}$:

Substitute $\dfrac{3R}{2}$ for s_1 in equation (2) and find that

$$s_1' = \frac{\left(\dfrac{3R}{2}\right)R}{2\left(\dfrac{3R}{2}\right) - R} = \frac{3R}{4}.$$

Now use this for s_1 and $\dfrac{3R}{2}$ for d in equation (4) and find that

$$s_2' = \frac{\left(\dfrac{3R}{2} - \dfrac{3R}{4}\right)R}{2\left(\dfrac{3R}{2} - \dfrac{3R}{4}\right) - R} = \frac{3R}{2}.$$

Since $d = \dfrac{3R}{2}$ is the vertex of mirror #1 in the second coordinate system, we've again shown that the image of the coin located at the vertex of one mirror (mirror #2) is at the vertex of the other mirror (mirror #1).

What about magnification? To find m, substitute the values we have found into equation (5).

$$m = \frac{s_2'}{s_1}\left(\frac{s_1'}{d - s_1'}\right) = \frac{\dfrac{3R}{2}}{\dfrac{3R}{2}}\left(\frac{\dfrac{3R}{4}}{\dfrac{3R}{2} - \dfrac{3R}{4}}\right) = 1.$$

Thus, the image is the same size as the original coin and is *not* inverted.

In terms of light rays, here's what happens. Rays from the coin travel to mirror #1. These are reflected back and form a *real* image half way between the two mirrors, at the point $\dfrac{3R}{4}$. This image is inverted. The rays continue towards mirror #2 as though they were coming from an ordinary inverted coin located halfway between the two mirrors. At mirror #2 they are reflected back towards mirror #1 and form a *real* image at mirror #1. This image is an inverted image of the first inverted image, and is therefore right side up.

23.49   Each lollipop forms a virtual image at a distance behind the mirror equal to its distance in front of the mirror. When they look in the mirror, the left-handed lollipops think that their images are right-handed, and the right-handed ones see their images as left-handed.

23.53

a)   Turn the diagram to put the problem into the standard geometry.

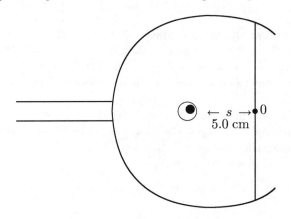

The refracting surface is *flat* with an infinite radius of curvature. Use the single surface refraction equation, with $s = -5.0$ cm.

$$-\frac{n_1}{s} + \frac{n_2}{s'} = \frac{n_2 - n_1}{R} \implies -\frac{1.33}{-5.0 \text{ cm}} + \frac{1.00}{s'} = \frac{1.00 - 1.33}{\infty \text{ cm}} = 0 \text{ cm}^{-1} \implies s' = -3.8 \text{ cm}.$$

The apparent depth of the olive is 3.8 cm. The olive appears 1.2 cm closer to the surface than it really is.

b) Use the single surface magnification equation.

$$m = \frac{n_1 s'}{n_2 s} = \frac{1.33(-3.8 \text{ cm})}{1.00(-5.0 \text{ cm})} = 1.0.$$

c) The image distance is negative, so the image is virtual. The magnification is positive, so the image is right-side-up.

23.57

a) and

b) Here's the picture.

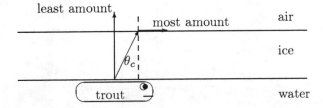

The light ray that spend the least amount of time in the ice is the ray that goes through the least amount of ice — the one that goes straight up.

The light ray that spends the greatest amount of time in the ice — but still (barely) gets through — is the ray that goes through the ice and strikes the ice-air interface at the critical angle θ_c to the normal.

c) Use the single surface refraction equation for a surface of infinite radius.

$$-\frac{n_1}{s} + \frac{n_2}{s'} = \frac{n_2 - n_1}{R} \implies \frac{1.30}{-1.00 \text{ m}} + \frac{1.00}{s'} = \frac{1.00 - 1.30}{\infty \text{ m}} = 0 \text{ m}^{-1} \implies s' = -0.77 \text{ m}.$$

The ice appears to be 0.77 m thick when viewed from above, even though its actual thickness is 1.00 m.

d) Use the single surface refraction equation for a surface of infinite radius, only now we go from ice to water.

$$-\frac{n_1}{s} + \frac{n_2}{s'} = \frac{n_2 - n_1}{R} \implies \frac{1.30}{-1.00 \text{ m}} + \frac{1.33}{s'} = \frac{1.33 - 1.30}{\infty \text{ m}} = 0 \text{ m}^{-1} \implies s' = -1.02 \text{ m}.$$

The ice appears to be 1.02 (m) thick when viewed by the fish, but it is actually 1.00 m thick.

Note that even though the fish's view is also distorted, it is distorted much less than ours. That is because the indices of refraction for water and ice are much closer together than the indices for air and ice.

23.61 For a real object, the object distance is always negative, so make this property explicit by writing

$$s = -|s|.$$

For a diverging lens, the focal length is negative, so make this explicit by writing

$$f = -|f|.$$

Write the thin lens equation with s and f in this form:

$$-\frac{1}{s} + \frac{1}{s'} = \frac{1}{f} \implies -\frac{1}{-|s|} + \frac{1}{s'} = \frac{1}{-|f|} \implies \frac{1}{s'} = -\frac{1}{|f|} - \frac{1}{|s|}.$$

This result means that $s' < 0$ m, so the image is always to the left of the lens in the standard geometry and is therefore virtual.

23.65

a)   The focal lengths of the two lenses are

$$2.0 \text{ dp} = \frac{1}{f_1} \implies f_1 = 0.50 \text{ m} = 50 \text{ cm} \qquad \text{and}$$

$$-5.0 \text{ dp} = \frac{1}{f_2} \implies f_2 = -0.20 \text{ m} = -20 \text{ cm}$$

Find the location s_1' of the image of the first lens relative to the first lens with the thin lens equation.

$$-\frac{1}{s_1} + \frac{1}{s_1'} = \frac{1}{f_1} \implies -\frac{1}{-150 \text{ cm}} + \frac{1}{s_1'} = \frac{1}{50 \text{ cm}} \implies s_1' = \frac{(50 \text{ cm})(150 \text{ cm})}{150 \text{ cm} - 50 \text{ cm}} = 75 \text{ cm}.$$

This image is 75 cm to the right of the first lens, and thus 25 cm to the right of the second. It is the object for the second lens, so $s_2 = +25 \text{ cm}$, a virtual object. Apply the thin lens equation to the second lens.

$$-\frac{1}{s_2} + \frac{1}{s_2'} = \frac{1}{f_2} \implies -\frac{1}{25 \text{ cm}} + \frac{1}{s_1'} = \frac{1}{-20 \text{ cm}} \implies s_1' = \frac{(20 \text{ cm})(25 \text{ cm})}{20 \text{ cm} - 25 \text{ cm}} = -1 \times 10^2 \text{ cm}.$$

The final image is 1×10^2 cm to the left of the diverging lens and 50 cm to the left of the converging lens.

b)   The magnification of the first lens is

$$m_1 = \frac{s_1'}{s_1} = \frac{75 \text{ cm}}{-150 \text{ cm}} = -0.50.$$

The magnification of the second lens is

$$m_2 = \frac{s_2'}{s_2} = \frac{-1 \times 10^2 \text{ cm}}{25 \text{ cm}} = -4.$$

The total magnification is the product of the individual magnifications.

$$m = m_1 m_2 = (-0.50)(-4) = 2.$$

c)   Since the final image distance is negative, the image is virtual. The total magnification is positive, so the final image is upright.

23.69   To find the focal length of the combination, place an object infinitely far from the lens. The location of the final image will by definition be the focal length of the combination.

First apply the thin lens equation to the object and the lens, in order to find the location s_{lens}' of the image formed by the lens.

$$-\frac{1}{s_{\text{lens}}} + \frac{1}{s_{\text{lens}}'} = \frac{1}{f_{\text{lens}}} \implies -\frac{1}{-\infty \text{ m}} + \frac{1}{s_{\text{lens}}'} = \frac{1}{f_{\text{lens}}} \implies s_{\text{lens}}' = f_{\text{lens}}$$

The image of the lens becomes the object for the mirror. Since the lens and mirror are in contact with each other, to a very good approximation the object distance for the mirror s_{mirror} is the image distance $s_{\text{lens}}' = f_{\text{lens}}$ of the lens. Apply the mirror equation to the mirror using $s_{\text{mirror}} = f_{\text{lens}}$.

$$\frac{1}{s_{\text{mirror}}} + \frac{1}{s_{\text{mirror}}'} = \frac{1}{f_{\text{mirror}}} \implies \frac{1}{f_{\text{lens}}} + \frac{1}{s_{\text{mirror}}'} = \frac{1}{f_{\text{mirror}}} \implies s_{\text{mirror}}' = \frac{f_{\text{lens}} f_{\text{mirror}}}{f_{\text{lens}} - f_{\text{mirror}}}.$$

This final image point is the focal length of the combination.

$$f_{\text{combination}} = \frac{f_{\text{lens}} f_{\text{mirror}}}{f_{\text{lens}} - f_{\text{mirror}}}.$$

23.73

a) and

b) The object (slide) and image (on the screen) are on opposite sides of the lens, so in the usual geometry s and s' have different signs: s' is positive, and s is negative. Therefore the magnification $m = \dfrac{s'}{s}$ is negative so the image is inverted.

The image of the largest side of the slide, the 3.40 cm side, will just fill the corresponding 150 cm side of the screen. Hence the absolute value of the magnification is

$$|m| = \frac{150 \text{ cm}}{3.40 \text{ cm}} = 44.1.$$

Therefore, $m = -44.1$. Insert the slide upside down and its image will have the proper orientation.

c) We are given the magnification

$$\frac{s'}{s} = m = -44.1 \implies s' = -44.1s$$

and the focal length $f = 10.0$ cm . Use these in the thin lens equation to solve for s.

$$-\frac{1}{s} + \frac{1}{s'} = \frac{1}{f'} \implies -\frac{1}{s} + \frac{1}{-44.1s} = \frac{1}{10.0 \text{ cm}} \implies -\frac{45.1}{44.1s} = \frac{1}{10.0 \text{ cm}} \implies s = -10.2 \text{ cm}.$$

The slide is 10.2 cm to the left (in the usual geometry) of the lens.

d) Use the magnification, $m = -44.1$, and object distance $s = -10.2$ cm , to find the image distance s'.

$$m = \frac{s'}{s} \implies s' = ms = (-44.1)(-10.2 \text{ cm}) = 450 \text{ cm} = 4.50 \text{ m}.$$

23.77 First determine the location of the images.

$$-\frac{1}{s} + \frac{1}{s'} = \frac{1}{f} \implies -\frac{1}{-50.0 \text{ cm}} + \frac{1}{s'} = \frac{1}{25.0 \text{ cm}} \implies s' = 50.0 \text{ cm}.$$

Next, determine the magnification of the images.

$$m = \frac{s'}{s} = \frac{50.0 \text{ cm}}{-50.0 \text{ cm}} = -1.00.$$

Therefore, the images are inverted and the same size as the objects. Here's the picture.

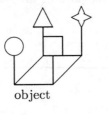
object

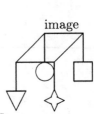
image

Note that the more distant objects are imaged *closer* to the lens.

23.81 Apply the thin lens equation to the first lens:

$$-\frac{1}{s} + \frac{1}{s'} = \frac{1}{f} \implies -\frac{1}{-30.0 \text{ cm}} + \frac{1}{s'} = \frac{1}{60 \text{ cm}} \implies s' = -60 \text{ cm}.$$

This image is virtual, is not inverted, and is on the same side of the lens as the object. The image of the first lens becomes the object for the second, so apply the thin lens equation to the second lens:

$$-\frac{1}{s} + \frac{1}{s'} = \frac{1}{f} \implies -\frac{1}{-60 \text{ cm}} + \frac{1}{s'} = \frac{1}{60 \text{ cm}} \implies s' = \infty \text{ cm}.$$

The image of the second lens becomes the object for the third, so apply the thin lens equation once again:

$$-\frac{1}{s} + \frac{1}{s'} = \frac{1}{f} \implies -\frac{1}{\infty \text{ cm}} + \frac{1}{s'} = \frac{1}{60 \text{ cm}} \implies s' = 60 \text{ cm}.$$

Thus, an object placed 30 cm to the left of this three-lens system will produce a real image 60 cm to the right of the system. (In case you are interested, since $m = \frac{s'}{s} = \frac{60 \text{ cm}}{-30 \text{ cm}} = -2.0$ is negative, the final image is inverted.)

23.85   To correct myopia, the focal length of the corrective eyeglass is the negative of the far point:

$$f = -d_{\text{far}} = -15 \text{ cm} = -0.15 \text{ m}.$$

The diopter value is the reciprocal of the focal length measured in meters. Hence,

$$\frac{1}{f} = \frac{1}{-0.15 \text{ m}} = -6.7 \text{ dp}.$$

23.89   Here's the geometry.

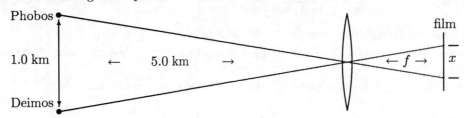

The object is effectively infinitely far from the lens, so the image is at the focal point of the camera lens (the picture greatly exaggerates the distance from lens to film compared to the distance from lens to mountains). The triangles on either side of the lens are similar, so

$$\frac{1.0 \text{ km}}{5.0 \text{ km}} = \frac{x}{f} \implies x = f\frac{1.0 \text{ km}}{5.0 \text{ km}} = (50.0 \text{ mm})\frac{1.0 \text{ km}}{5.0 \text{ km}} = 10 \text{ mm}.$$

23.93

a)   After a few simple measurements, the results for one of the authors are shown below:

The arm is the 62 cm radius of a circle. The hand takes up about 7.5 cm of this circumference of the circle. Hence the hand fills an angle of about

$$\frac{7.5 \text{ cm}}{62 \text{ cm}} \text{ rad} = 0.12 \text{ rad} = (0.12 \text{ rad})\left(\frac{2\pi \text{ rad}}{360°}\right) = 6.9°.$$

b)   The angular size of the Moon is

$$\frac{3.48 \times 10^6 \text{ m}}{3.84 \times 10^8 \text{ m}} = 9.06 \times 10^{-3} \text{ rad} = (9.06 \times 10^{-3} \text{ rad})\left(\frac{360°}{2\pi \text{ rad}}\right) = 0.519°.$$

c)   The focal length of the mirror is half the radius of curvature. Since the mirror is concave, the focal length is negative.

$$f = -12.00 \text{ m}.$$

d)   This angle is the same is in part (c),

$$9.06 \times 10^{-3} \text{ rad} = 0.519°.$$

e)   Since the distance to the Moon is many times the focal length of the mirror, it is effectively an infinite distance away. Therefore, the image of the Moon is at the focal point of the concave mirror, and is a real image.

f)   Let x be the linear width of the image of the Moon. Here's the geometry.

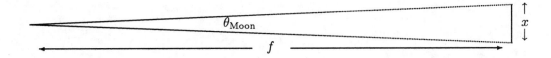

$$\theta_{\text{Moon}} = \frac{x}{|f|} \implies x = \theta_{\text{Moon}}|f| = (9.06 \times 10^{-3} \text{ rad})(12.00 \text{ m}) = 0.109 \text{ m}.$$

g)   The image of the Moon is circular. The shape of the image depends upon the shape of the object, not upon the shape of the perimeter of the optical element forming the image.

23.97

a)   The magnification is

$$m = -\frac{f_o}{f_e} = -\frac{2.0 \text{ cm}}{20.0 \text{ cm}} = -0.10.$$

The image subtends an angle one tenth that subtended by the object, and it is upside down.

b)   If we place the object close to but outside of the secondary focal point of the objective lens, then the system acts as a *microscope*. The magnification is

$$m \approx -\frac{(25 \text{ cm})L}{f_o f_e} = \frac{(25 \text{ cm})(22 \text{ cm})}{(2.0 \text{ cm})(20.0 \text{ cm})} = -14.$$

Chapter 24

Physical Optics

24.1   Use the equation for locating the maxima for double slit interference:

$$d\sin\theta = m\lambda.$$

The zero order fringe has $m = 0$, so

$$d\sin\theta_0 = 0\lambda = 0\text{ m} \implies \theta_0 = 0\text{ rad} = 0°.$$

The first order fringe has $m = 1$, for it we have

$$d\sin\theta_1 = 1\lambda \implies (1.4 \times 10^{-4}\text{ m})\sin\theta_1 = 488 \times 10^{-9}\text{ m} \implies \sin\theta_1 = 3.5 \times 10^{-3}.$$

If you use your calculator to find θ_1, you'll find that $\theta_1 = \arcsin(3.5 \times 10^{-3}) = 0.0035000071$ rad. Notice that this is the same as θ_1 to *five* significant digits, and differs from θ_1 by less than one in the sixth significant digit! Since the data of the problem only entitle us to keep two significant digits, we may regard the small angle "approximation" $\sin\theta \approx \theta$ as being exact for all practical purposes, that is, for the size of angles and amount of precision we are dealing with in most problems of this type.

$$\theta_1 = 3.5 \times 10^{-3}\text{ rad} = (3.5 \times 10^{-3}\text{ rad})\left(\frac{360°}{2\pi}\right) = 0.20°.$$

The angle between the zero and first order fringe is $\theta_1 - \theta_0 = 3.5 \times 10^{-3}$ rad $= 0.20°$.

24.5

a)   The angular position θ of the first order fringe is found from the equation locating the maxima of the double slit interference pattern:

$$d\sin\theta = m\lambda \implies (0.10 \times 10^{-3}\text{ m})\sin\theta = 1(632.8 \times 10^{-9}\text{ m})$$

$$\implies \sin\theta = \frac{632.8 \times 10^{-9}\text{ m}}{0.10 \times 10^{-3}\text{ m}} = 6.3 \times 10^{-3} \implies \theta = 6.3 \times 10^{-3}.$$

Note, for angles this small, $\sin\theta$ and θ (in radians) are the same to within *five* significant digits, so, since the other data are accurate to within only two or three significant digits, the small angle "approximation" $\sin\theta \approx \theta$, is for all practical purposes an equality.

More generally, the separation between any two adjacent maxima, m, and $m - 1$, is calculated the same way. For small angles, the angular separation is

$$\theta_m - \theta_{m-1} \approx \sin\theta_m - \sin\theta_{m-1} = \frac{m\lambda}{d} - \frac{(m-1)\lambda}{d} = \frac{\lambda}{d}.$$

233

Again, the approximation is so good, that we may treat it as an equality, so

$$\theta_m - \theta_{m-1} = \frac{\lambda}{d} = \frac{632.8 \times 10^{-9} \text{ m}}{0.10^{-3} \text{ m}} = 6.3 \times 10^{-3} \text{ rad}.$$

Now let x_m be the separation between the m and $m-1$ fringes on the screen, and let s be the distance from the slits to the screen. Here's the picture for $m = 1$ and $m = 2$.

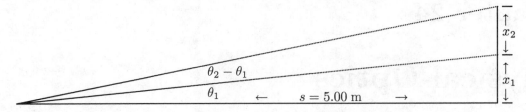

Although, strictly speaking, from the above geometry,

$$x_m = s \sin \theta_m - s \sin \theta_{m-1},$$

because the angles θ_m are so small, we may treat the approximation $\sin \theta_m \approx \theta_m$ as an equality and whatever error there is in this is far less than the errors in measurement. Hence, well within the number of displayed significant figures, we have

$$x_m = s\theta_m - s\theta_{m-1} = s\frac{\lambda}{d} = (5.00 \text{ m})(6.3 \times 10^{-3} \text{ rad}) = 3.2 \times 10^{-2} \text{ m} = 3.2 \text{ cm}.$$

b)   From the equation $\theta_m - \theta_{m-1} = \dfrac{\lambda}{d}$, we see that the bigger the slit separation d, the closer the fringes are to each other.

24.9

a)   The wavelength λ of the sound waves is found from

$$v = \nu\lambda \implies \lambda = \frac{v}{\nu} = \frac{343 \text{ m/s}}{262 \text{ Hz}} = 1.31 \text{ m}.$$

b)   Use the equation for the double slit interference maxima with $m = 1$:

$$d \sin \theta = m\lambda \implies \sin \theta = m\frac{\lambda}{d} = 1\frac{1.31}{2.0} \implies \theta = 0.71 \text{ rad} = 41°.$$

Locate yourself 41° on either side of the straight through direction to hear Pavoratti loud and clear.

24.13

a)   Use Equation 24.5 on page 1111 of the text to locate the angular position θ of the first minimum of a circular diffraction pattern.

$$a \sin \theta = 1.220\lambda \implies a = \frac{1.220\lambda}{\sin \theta} = \frac{1.220(600 \times 10^{-9} \text{ m})}{\sin(1.00 \times 10^{-3} \text{ rad})} = 7.3 \times 10^{-4} \text{ m} = 0.73 \text{ mm}.$$

So, the diameter of the circular hole is 0.73 mm.

b)   For a given wavelength λ, the right-hand side of

$$a \sin \theta = 1.220\lambda$$

is a constant. Hence, if you increase the size a of the aperture, $\sin \theta$ must decrease to compensate, and thus the size of the central diffraction pattern becomes smaller.

24.17 The angular width of the 5 m wall when viewed from the moon, 3.63×10^8 m away, is

$$\theta = \frac{5 \text{ m}}{3.63 \times 10^8 \text{ m}} = 1 \times 10^{-8} \text{ rad}.$$

The aperture of the eye is about 1 cm. The Rayleigh criterion says that the angular resolution limit θ_{eye} of the eye for light with a wavelength of 500 nm satisfies

$$a \sin \theta_{eye} = 1.220\lambda \implies \sin \theta_{eye} = \frac{1.220\lambda}{a} = 6 \times 10^{-5} \implies \theta_{eye} = 6 \times 10^{-5} \text{ rad}.$$

The angular width of the wall is far less than the resolution limit of the eye, so the astronauts cannot see the wall.

It is also interesting to modify these calculations, and calculate just how wide a wall on the Earth must be to be visible to the naked eye of an astronaut on the Moon. Use the angular resolution of the eye to find the smallest width x of a wall that can be seen:

$$\frac{x}{3.63 \times 10^8 \text{ m}} = \theta_{eye} \implies x = (3.63 \times 10^8 \text{ m})\theta_{eye}$$

$$= (3.63 \times 10^8 \text{ m})(6 \times 10^{-5} \text{ rad}) = 2 \times 10^4 \text{ m} = 20 \text{ km} \approx 13 \text{ mile}!$$

Of course, we get the same answer if we ask what the minimum width we on Earth can resolve for objects on the Moon. If someone tells you they see astronauts on the Moon, they must be very big astronauts!

24.21

a) The larger the aperture, the better the better resolution, so details will be better seen through the telescope.

b) Use the Rayleigh criterion (for a circular aperture) to determine the *angular* separation of two barely resolved objects:

$$a \sin \theta = 1.220\lambda \implies \sin \theta = 1.220\frac{\lambda}{a}.$$

For small angles, $\sin \theta \approx \theta$, so

$$\theta = 1.220\frac{\lambda}{a} = 1.220\left(\frac{550 \times 10^{-9} \text{ m}}{10 \times 10^{-2} \text{ m}}\right) = 6.7 \times 10^{-6} \text{ rad}.$$

Now let x be the *distance* separating the objects on the mountain, and let the distance of the mountain be d. If they can just barely be resolved, then their angular separation is θ, so

$$\theta = \frac{x}{d} \implies x = \theta d = (6.7 \times 10^{-6} \text{ rad})(150 \times 10^3 \text{ m}) = 1.0 \text{ m}.$$

So the telescope can resolve objects separated by one meter or more.

c) The magnification of a telescope is given by the ratio of the focal lengths of the objective and eyepieces:

$$m = \frac{f_o}{f_e}.$$

If you halve the focal length of the eyepiece, you double the magnification. Since the size of the objective is the same, there is no effect upon the resolution. That is, the smallest separation that can be resolved is still 1.0 m.

24.25   Use the grating equation, Equation 24.8 on page 1116 of the text:

$$d \sin \theta_m = m\lambda \implies \sin \theta_m = m\frac{\lambda}{d}.$$

For the first order, $m = 1$, so

$$\sin \theta_1 = 1 \left(\frac{488 \times 10^{-9} \text{ m}}{2.0 \times 10^{-6} \text{ m}} \right) = 0.24 \implies \theta_1 = 0.25 \text{ rad} = 14°.$$

For the second order, $m = 2$, so

$$\sin \theta_2 = 2 \left(\frac{488 \times 10^{-9} \text{ m}}{2.0 \times 10^{-6} \text{ m}} \right) = 0.49 \implies \theta_1 = 0.51 \text{ rad} = 29°.$$

24.29   In order to use the grating equation, first find the distance d between slits.

$$d = \frac{1}{1.000 \times 10^4} \text{ cm} = 1.000 \times 10^{-4} \text{ cm} = 1.000 \times 10^{-6} \text{ m}.$$

Now use the grating equation for $m = 1$.

$$d \sin \theta = m\lambda \implies \sin \theta = \frac{\lambda}{d}.$$

At the 400 nm end we have

$$\sin \theta_{400} = \frac{400 \times 10^{-9} \text{ m}}{1.000 \times 10^{-6} \text{ m}} = 0.400 \implies \theta_{400} = 0.412 \text{ rad} = 23.6°.$$

At the 700 nm end we have

$$\sin \theta_{700} = \frac{700 \times 10^{-9} \text{ m}}{1.000 \times 10^{-6} \text{ m}} = 0.700 \implies \theta_{700} = 0.775 \text{ rad} = 44.4°.$$

The angular spread of the first order spectrum is the difference in these angles,

$$0.775 \text{ rad} - 0.412 \text{ rad} = 0.363 \text{ rad} = 20.8°.$$

24.33

a)   and

b)   The second order peak for 656 nm has the same angle as the third order peak for the unknown wavelength λ_{unknown}. Hence, in the grating equation, $d \sin \theta = m\lambda$, both d and $\sin \theta$ are the same, so

$$2(656 \text{ nm}) = d \sin \theta = 3(\lambda_{\text{unknown}}) \implies \lambda_{\text{unknown}} = \frac{2}{3}(656 \text{ nm}) = 437 \text{ nm}.$$

So it was not necessary to know the grating spacing d, and we don't get the rest of the day off, Ψ! (sigh!).

24.37

a)   Use Equation 24.20 on page 1121 of the text, which gives the wavelength λ_1 of light in a medium other than a vacuum as a function of the index of refraction n_1 of that medium and the wavelength λ in a vacuum:

$$\lambda_1 = \frac{\lambda}{n_1} = \frac{632.8 \text{ nm}}{1.50} = 422 \text{ nm}.$$

b)   Use Equation 24.17 on page 1120 of the text, which relates the speed c of light in a vacuum to its speed v and the index of refraction n of a medium other than a vacuum:

$$n = \frac{c}{v} \implies v = \frac{c}{n} = \frac{3.00 \times 10^8 \text{ m/s}}{1.50} = 2.00 \times 10^8 \text{ m/s}.$$

24.41 We'll assume internal reflections are negligible.

a) The amount of time a ray spends in the sphere, is directly proportional to the length of the path that it takes through the sphere. the longest paths through the sphere are those that pass directly through the center of the sphere. these are the ones that traverse one of the sphere's diameters.

Any entering ray whose incident angle is 0° (to the normal) will continue, by the law of refraction, through the sphere at a 0° angle. Hence, any such ray will stay inside the sphere as long as possible.

b) The length of time for the light to traverse the sphere is the distance d traveled divided by the speed v. The distance is the diameter $d = 2.00$ cm $= 2.00 \times 10^{-2}$ m Use the index of refraction n to determine v.

$$n = \frac{c}{v} \implies v = \frac{c}{n} = \frac{3.00 \times 10^8 \text{ m/s}}{1.50} = 2.00 \times 10^8 \text{ m/s}.$$

So,

$$t_{max} = \frac{2.00 \times 10^{-2} \text{ m}}{2.00 \times 10^8 \text{ m/s}} = 1.00 \times 10^{-10} \text{ s} = 0.100 \text{ ns}.$$

c) At first it might seem that the incident rays that spend the least time in the sphere are those whose angle of incidence is 90°, and that these rays leave as soon as they enter and therefore spend zero time in the sphere.

It is correct that these are the minimum-time rays. But such rays *don't spend zero time in the sphere.* The reason for this is that they enter the sphere with an angle of refraction equal to the critical angle, and after that they travel a nonzero distance through the sphere.

In order to be careful about this, let's look at an *arbitrary* ray whose angle of incidence is θ_1, and find out how far it travels in the sphere. We can then find the value of θ_1 that minimizes this distance, and therefore the time. Here's the picture.

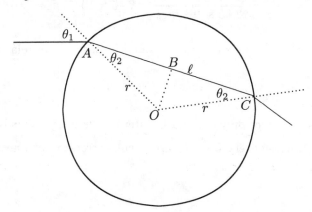

The ray enters the sphere at the point A at an angle θ_1 to the normal. It is then refracted at an angle θ_2, travels a distance ℓ (yet to be determined) through the sphere, and exits the sphere at the point C. We have drawn a point B is halfway from A to C. We also have drawn lines OA, OB, and OC from the center of the sphere to the points A, B, and C. The triangles ABO and CBO have corresponding sides equal to each other and are therefore congruent. Therefore $\angle OCB = \theta_2$, as marked, and $\angle OBA = \angle OBC = 90°$

From the above geometry, $\frac{\ell}{2} = r \cos\theta_2$, so

$$\ell = 2r\cos\theta_2.$$

We want to rewrite this expression in terms of θ_1. By the law of refraction, $n_1 \sin \theta_1 = n_2 \sin \theta_2$, where n_1 and n_2 are the indices of refraction. In this case, $n_1 = 1.00$, and $n_2 = 1.50$. Therefore

$$\sin \theta_2 = \frac{1}{1.50} \sin \theta_1 \implies \cos \theta_2 = \sqrt{1 - \left(\frac{1}{1.50} \sin \theta_1 \right)^2},$$

so

(1)
$$\ell = 2r \sqrt{1 - \left(\frac{1}{1.50} \sin \theta_1 \right)^2}.$$

This expression is at a *maximum* when $\sin \theta_1 = 0 \implies \theta_1 = 0°$. This agrees with our answer to part a).

The expression for ℓ is at a *minimum* when $\sin \theta_1 = 1 \implies \theta_1 = 90°$. So the incident rays that spend minimum time in the sphere are those whose angle of incidence is 90°. These rays arrive tangent to the surface of the sphere, but are bent by the higher index of refraction of the glass and end up going a distance $\ell > 0$ m through the sphere.

d)   To calculate the *minimum* time, first use equation (1) above with $\theta_1 = 90°$ to find the minimum distance ℓ_{min}, and then use this minimum distance to find the minimum time.

From equation (1) with $\theta_1 = 90°$ we have

$$\ell_{min} = 2r \sqrt{1 - \left(\frac{1}{1.50} \sin 90° \right)^2} = (2.00 \times 10^{-2} \text{ m}) \sqrt{1 - \left(\frac{1}{1.50} \right)^2} = 1.49 \times 10^{-2} \text{ m}.$$

In part a) we found the speed v of light inside the sphere is $v = 2.00 \times 10^8$ m/s. Therefore the time required to travel the minimum distance is

$$t_{min} = \frac{\ell_{min}}{v} = \frac{1.49 \times 10^{-2} \text{ m}}{2.00 \times 10^8 \text{ m/s}} = 7.45 \times 10^{-11} \text{ s} = 0.0745 \text{ ns}.$$

This is almost three quarters of the maximum time found in part a), so, relatively speaking, it is far from zero!

24.45

a)   The one labeled #1 is flat, since only one (dark) interference fringe is visible.

b)   Since the interference fringes observed in #2 are circular, tracing out lines of equal thickness, # 2 has a slightly spherical shape.

24.49

a)   The ray reflected from the glass to air interface experiences no phase change upon reflection. However, the ray reflected from the air to glass interface does experience a phase change of π rad upon reflection. Hence, the condition for a bright fringe is that the path difference $2d$ must be an odd number of half wavelengths (in air, since the "thin film" is an air film), so

$$2d = (2m + 1)\frac{\lambda}{2}.$$

Here's the picture.

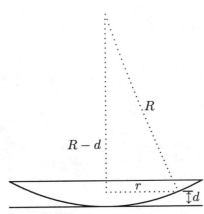

By the Pythagorean theorem, $r = \sqrt{R^2 - (R-d)^2} = \sqrt{2Rd - d^2}$. As a part of the "thin film approximation," d is small, so d^2 is very small, and therefore

(1) $$r \approx \sqrt{2Rd}.$$

Since $2d = (2m+1)\frac{\lambda}{2}R$, this implies

$$r \approx \sqrt{(2m+1)\frac{\lambda}{2}R}.$$

b)   For a dark ring, the path difference must be an integral number of wavelengths,

$$2d = m\lambda.$$

Use this in (1) above:

$$r \approx \sqrt{m\lambda R}.$$

24.53   The reflected light is completely plane polarized when its angle of incidence is the Brewster angle. Here's the picture.

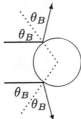

According to Equation 24.31 on page 1131 of the text, the Brewster angle satisfies

$$\tan\theta_B = \frac{n_2}{n_1} = \frac{1.50}{1.00} = 1.50 \implies \theta_B = 0.983 \text{ rad} = 56.3°.$$

As shown in the above drawing, the reflected ray makes an angle $2\theta_B = 112.6°$ with the incident ray.

24.57

a)   and

b)   The tangent function is an increasing function and $\tan 45° = 1$. Thus,

$$n_2 > n_1 \implies \frac{n_2}{n_1} = \tan\theta_B > 1 \implies \theta_B > 45°,$$

and

$$n_2 < n_1 \implies \frac{n_2}{n_1} = \tan\theta_B < 1 \implies \theta_B < 45°.$$

c)   Let light inside the material with an index of refraction of 4.5 reflect off the material to air interface. In this case, the Brewster angle satisfies

$$\tan \theta_B = \frac{n_2}{n_1} = \frac{1.00}{4.50} = 0.222 \implies \theta_B = 0.218 \text{ rad} = 12.5°.$$

So this would be the smallest Brewster angle one would ever find in the laboratory.

Using the *Handbook of Chemistry and Physics*, the highest n we can find is the index of refraction for galenite (PbS), which is $n = 3.9$ when λ is an *optical* wavelength . Aluminum has an index of refraction of about 4.5 when $\lambda \approx 3.1 \times 10^{-6}$ m (i.e., when λ is in the infrared).

The critical angle θ_c for such an interface satisfies

$$4.5 \sin \theta_c = 1.00 \sin 90° \implies \sin \theta_c = 0.222 \implies \theta_c = 0.224 \text{ rad} = 12.8°.$$

The critical angle is slightly greater than the Brewster angle, so the light is not totally internally reflected when incident at the Brewster angle.

24.61   When the light is incident at the Brewster angle, all of the reflected light will be polarized, so

$$\tan \theta_B = \frac{n_2}{n_1} = \frac{1.33}{1.00} = 1.33 \implies \theta_B = 0.926 \text{ rad} = 53.1°.$$

This is the angle that the Moon makes with the normal to the lake. The elevation of the Moon is the angle that it makes to the horizon, and is therefore the complement of this angle.

$$\text{elevation of Moon} = 90° - 53.1° = 36.9°.$$

So if you are watching the reflection of the rising Moon in the lake with your Polaroid® sunglasses on, you will see it grow fainter and fainter until it is at an angle of 36.9°, at which point it will disappear. As the Moon rises to higher elevations its reflection will gradually reappear.

Chapter 25

The Special Theory of Relativity

25.1 Measure the distance from one clock to another. Calculate the time for light to travel from a given clock to another and set the latter clock accordingly to compensate for the travel time of the light.

For example: Arrange matters so that each clock can sense a single light signal that you are going to send from the origin, and each one will start up as soon as it gets the signal.

Now go around and place the clocks, and set the time on each one. Set the clocks — but don't start them! Here's how you set them. For any given clock set the time on the clock equal to $\dfrac{d}{c}$, where d is the distance of the clock from the origin, and c is the speed of light.

Once you have all the clocks in place and all *set* (but not started), go back to the origin and send the common light signal. As soon as a clock senses the light signal, it will start. Clocks farther from the origin will start later — but that's alright, you've set their initial times to compensate for this.

25.5 We are told that when 1.00 s in the rest frame of the clock, 3.00 s passes in our frame. We want to find the speed v of the clock's frame relative to ours. Use the equation for time dilation.

$$\tau = \gamma \tau_0 \implies 3.00\text{ s} = \gamma(1.00\text{ s}) \implies \gamma = 3.00 \implies \frac{1}{\sqrt{1 - \dfrac{v^2}{c^2}}} = 3.00.$$

Square both sides of the last equation and solve for v.

$$\frac{1}{1 - \dfrac{v^2}{c^2}} = 9.00 \implies 1 - \frac{v^2}{c^2} = \frac{1.00}{9.00} \implies \frac{v^2}{c^2} = \frac{8.00}{9.00}$$

$$\implies v = c\sqrt{\frac{8.00}{9.00}} = (3.00 \times 10^8 \text{ m/s})\sqrt{\frac{8.00}{9.00}} = 2.83 \times 10^8 \text{ m/s}.$$

25.9 Use the calculus to linearly approximate $y = x^2$ near $x = 1$.

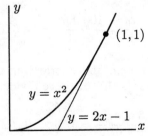

The approximation is

$$x^2 \approx 2x - 1.$$

241

Now use this expression to approximate $\dfrac{v^2}{c^2}$ in the formula for γ.

$$\gamma = \frac{1}{\sqrt{1-\dfrac{v^2}{c^2}}} \approx \sqrt{\frac{1}{1-\left(2\dfrac{v}{c}-1\right)}} = \sqrt{\frac{1}{2\left(1-\dfrac{v}{c}\right)}}.$$

For $\dfrac{v}{c} = 0.999\,999$, the exact expression for γ gives

$$\gamma = 707.1070$$

to 7 significant figures. The approximation gives

$$\gamma \approx 707.1068,$$

to 7 significant figures. So our approximation works well when $\dfrac{v}{c}$ is very close to 1.

   Approximations such as this were commonly used before the advent of electronic calculators (pre 1970s).

25.13

a)   Your roommate calculates your speed by dividing the distance traveled by the time as measured by his rulers and clocks at rest in his reference frame. So,

$$v = \frac{100.0 \text{ m}}{5.00 \times 10^{-7} \text{ s}} = 2.00 \times 10^8 \text{ m/s} = \frac{2}{3}c.$$

b)   and

c)   In your reference frame you are at rest, and the corridor is moving past you at a speed $v = \dfrac{2}{3}c$. Consequently, relative to you, the length of the corridor is

$$\ell = \frac{\ell_0}{\gamma} = \frac{100.0 \text{ m}}{\gamma}.$$

So, to find ℓ we need first to find γ:

$$\gamma = \frac{1}{\sqrt{1-\dfrac{v^2}{c^2}}} = \frac{1}{\sqrt{1-\dfrac{(2.00 \times 10^8 \text{ m/s})^2}{(3.00 \times 10^8 \text{ m/s})^2}}} = 1.34.$$

Therefore, relative to you, the length of the moving corridor is

$$\ell = \frac{\ell_0}{\gamma} = \frac{100.0 \text{ m}}{1.34} = 74.6 \text{ m}.$$

According to you, the corridor is 74.6 m long and is traveling past you at a speed of 2.00×10^8 m/s. Therefore, you say the time it takes to pass is this distance divided by the speed:

$$t' = \frac{74.6 \text{ m}}{2.00 \times 10^8 \text{ m/s}} = 3.73 \times 10^{-7} \text{ s} = 373 \text{ ns}.$$

25.17

a) Assume that α-Centauri and the Earth are at rest relative to one another. You see yourself at rest, α-Centauri approaching you at a speed v, and the Earth leaving you at a speed v. Therefore, relative to you, the distance from Earth to α-Centauri is contracted, and is

$$\ell = \frac{\ell_0}{\gamma} = \frac{4.2 \text{ LY}}{\gamma}.$$

In your reference frame, it takes you 3.0 y to travel this distance. Therefore, in your reference frame,

$$\text{distance} = (\text{rate})(\text{time}) \implies \ell = v(3.0 \text{ y}) \implies \frac{4.2 \text{ LY}}{\gamma} = v(3.0 \text{ y}) \implies \gamma v = \frac{4.2 \text{ LY}}{3.0 \text{ y}} = 1.4 \left(\frac{\text{LY}}{\text{y}} \right).$$

Now one light year per year is just the speed c of light. So, $1.4 \left(\dfrac{\text{LY}}{\text{y}} \right) = 1.4c$. Thus

$$\gamma v = 1.4c.$$

We want to solve this equation for v. Since γ also depends upon v, we rewrite γ in terms of v and turn the proverbial mathematical crank:

$$\gamma v = 1.4c \implies \frac{1}{\sqrt{1 - \dfrac{v^2}{c^2}}} v = 1.4c \implies \frac{v^2}{1 - \dfrac{v^2}{c^2}} = (1.4)^2 c^2$$

$$\implies \frac{v^2}{c^2} = (1.4)^2 - (1.4)^2 \frac{v^2}{c^2}$$

$$\implies \frac{v^2}{c^2} = \frac{(1.4)^2}{1 + (1.4)^2}$$

$$\implies \frac{v}{c} = 0.81$$

So $v = 0.81c = 0.81(3.00 \times 10^8 \text{ m/s}) = 2.4 \times 10^8 \text{ m/s}$.

b) The distance in your friends' reference frame is 4.2 LY and they see you traveling at a constant speed of $0.81c$. Hence, in their reference frame, the time required for the journey is

$$\frac{4.2 \text{ LY}}{0.81c} = 5.1 \left(\frac{\text{LY}}{c} \right) = 5.1 \text{ y}.$$

25.21

a) The time Δt required is the distance divided by the speed. So,

$$\Delta t = \frac{3.84 \times 10^8 \text{ m}}{3.00 \times 10^8 \text{ m/s}} = 1.28 \text{ s}.$$

It takes 1.28 s for light to get here from the Moon.

b) Call the time at which the meteor crashes into the lecture room $t = 0$ s. Astronomers at the college receive light from the Moon at time $t = 0 \text{ s} + 1.10 \text{ s} = 1.10 \text{ s}$. However, from part a), it took 1.28 s for the light to get here from the Moon. Therefore, in our frame, the meteor must have hit the Moon at time $t = -0.18$ s. Thus, in the reference frame of the lecture room, the Moon was hit by its meteor 0.18 s *before* the lecture hall was hit by its meteor.

c) According to part b), the first meteor was the one that hit the Moon. Therefore, if the meteors were dropped by the *same* flying saucer, the saucer would have had to drop its meteor on the Moon and then race to get here and drop the second meteor on us before the light from the first meteor strike got here. Since we don't believe material objects can travel faster than light, this is impossible. If the meteor strikes were due to flying saucers, there must have been more than one saucer.

(In the theory of Special Relativity, one can prove that any massive object with $v < c$ can never attain speeds $v \geq c$ because to do so would require an infinite amount of energy. One also can prove that if any information travels with $v > c$ causality will be violated. That is, if event A happens before event B in one inertial frame, then there exist other inertial frames in which event B happens before event A. Since causality violations have never been observed, and since assuming $v > c$ predicts causality violations, we believe that whenever Special Relativity is valid no information can travel faster than the speed of light.)

25.25

a)   Length contraction is expressed quantitatively by the equation

$$\ell = \frac{\ell_0}{\gamma} = \frac{1}{\left(\sqrt{1 - \left(\frac{v^2}{c^2}\right)}\right)^{-1}} = \ell_0 \sqrt{1 - \left(\frac{v^2}{c^2}\right)}.$$

We'd like to approximate this by some linear function of $\frac{v^2}{c^2}$, so for the time being let $x = \frac{v^2}{c^2}$, and write

$$\ell(x) = \ell_0 \sqrt{1 - x}.$$

Now use the calculus to linearly approximate $\ell(x)$ near $x = 0$. The derivative of $\ell(x)$ is $\frac{d\ell}{dx} = -\frac{\ell_0}{2\sqrt{1 - x}}$, so at $x = 0$ the slope of our approximation is $-\frac{\ell_0}{2\sqrt{1 - 0}} = -\frac{\ell_0}{2}$. The vertical intercept at $x = 0$ is ℓ_0. Here's the picture.

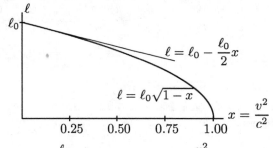

The linear approximation is $\ell_{\text{approx}} = \ell_0 - \frac{\ell_0}{2}x$, so, replacing x by $\frac{v^2}{c^2}$,

$$\ell \approx \ell_{\text{approx}} = \ell_0 - \frac{\ell_0}{2}\frac{v^2}{c^2} \implies \Delta\ell \approx -\frac{\ell_0}{2}\frac{v^2}{c^2}.$$

This approximation is quite good even for speeds as high as 25% the speed of light: When $\frac{v}{c} = 0.250$, we have $\ell = 0.9682\ell_0$ and $\ell_{\text{approx}} = 0.9688\ell_0$. For $\frac{v}{c} = 0.100\,00$, we have $\ell = 0.994\,99\ell_0$ and $\ell_{\text{approx}} = 0.995\,00\ell_0$.

We used the calculus, but the problem statement suggests using the "binomial expansion," to find the approximation. For those of you not familiar with the binomial expansion, here is a brief exposition:

The *binomial expansion* for $(1 + x)^n$ is

(1)
$$(1 + x)^n = \sum_{k=0}^{k=\infty} \binom{n}{k} x^k.$$

Where the coefficients $\binom{n}{k}$ are called (strangely enough!) "the binomial coefficients." They are given by

$$\binom{n}{k} = \begin{cases} \dfrac{n(n - 1)(n - 2)\cdots(n - k + 1)}{k(k - 1)(k - 2)\cdots 1} & \text{for } k > 0 \\[2mm] 1 & \text{for } k = 0. \end{cases}$$

25.29.

When n is a positive integer, the binomial coefficients are 0 for $k > n$, and the sum in (1) is a finite sum. In this case, equation (1) is often proved in high school, and the binomial coefficients can also be found from "Pascal's triangle." In the case where n is not a positive integer, the sum in (1) is in fact an infinite power series whose radius of convergence is equal to 1. In this case, the sum is the Taylor series expansion for $(1 + x)^n$.

In the particular case where $n = \dfrac{1}{2}$, the first few binomial coefficients are

$$\binom{1/2}{0} = 1, \quad \binom{1/2}{1} = \frac{1}{2}, \quad \binom{1/2}{2} = -\frac{1}{8}, \quad \text{and} \quad \binom{1/2}{3} = \frac{1}{16}.$$

So,

$$\sqrt{1 - x} = (1 + [-x])^{1/2} = 1 + \frac{1}{2}[-x] - \frac{1}{8}[-x]^2 + \frac{1}{16}[-x]^3 \cdots$$

$$= 1 - \frac{1}{2}x - \frac{1}{8}x^2 - \frac{1}{16}x^3 \cdots$$

A linear approximation to $\sqrt{1 - x}$ is obtained by neglecting all but the first two terms in this sum — which is all right provided that $|x| << 1$. When we do this, we obtain the same approximation as we did above:

$$\sqrt{1 - x} \approx 1 - \frac{1}{2}x.$$

So,

$$\ell = \ell_0 \sqrt{1 - \frac{v^2}{c^2}} \approx \ell_0 \left(1 - \frac{1}{2}\frac{v^2}{c^2}\right) \implies \Delta\ell \approx -\frac{1}{2}\frac{v^2}{c^2}\ell_0.$$

b)   Use the approximation from part a).

$$\Delta\ell \approx -\frac{1}{2}\frac{v^2}{c^2}\ell_0 = -\frac{1}{2}\frac{(30 \times 10^3 \text{ m/s})^2}{(3.00 \times 10^8 \text{ m/s})^2}(12.7 \times 10^6 \text{ m}) = -0.064 \text{ m}.$$

Only the diameter (or the component of the diameter) in the direction of the Earth's orbital motion is contracted.

25.29   Here is a picture of the situation, viewed from S.

a)   Use the relativistic velocity component addition equation:

$$u_x = \frac{u_{x'} + v}{1 + \dfrac{u_{x'}v}{c^2}} = \frac{(0 \text{ m/s}) + v}{1 + \dfrac{(0 \text{ m/s})v}{c^2}} = v = 0.990c.$$

For the other velocity component, use

$$u_y = \frac{u_{y'}}{\gamma\left(1 + \dfrac{vu_{x'}}{c^2}\right)} = \frac{u_{y'}}{\gamma\left(1 + \dfrac{v(0 \text{ m/s})}{c^2}\right)} = \frac{u_{y'}}{\gamma}.$$

Now we need the numerical value of γ.

$$\gamma = \frac{1}{\sqrt{1 - \dfrac{v^2}{c^2}}} = \frac{1}{\sqrt{1 - (0.990)^2}} = 7.1$$

Hence

$$u_y = \frac{u_{y'}}{\gamma} = \frac{0.995c}{7.1} = 0.14c.$$

b)   The angle ϕ that the velocity $\vec{u}$ makes with the x-axis in S is found from

$$\tan \phi = \frac{u_y}{u_x} = \frac{0.14c}{0.990c} = 0.14 \implies \phi = 0.14 \text{ rad } = 8.0°.$$

25.33   Here are the two equations we need to derive:

(a_x)

$$a_x = \frac{a_{x'}}{\gamma^3 \left(1 + \dfrac{u_{x'}v}{c^2}\right)^3},$$

and

(a_y)

$$a_y = \frac{\left(1 + \dfrac{vu_{x'}}{c^2}\right) a_{y'} - u_{y'} \dfrac{va_{x'}}{c^2}}{\gamma^2 \left(1 + \dfrac{vu_{x'}}{c^2}\right)^3}$$

We'll begin with a_x. We'll find it by differentiating the transformation equation for the x-component of the velocity, which is Equation 25-45 on page 1172 of the text:

(1)

$$u_x = \frac{u_{x'} + v}{1 + \dfrac{u_{x'}v}{c^2}}.$$

Differentiate u_x with respect to t using the chain rule, the quotient rule, and the relation $\dfrac{du_{x'}}{dt'} = a_{x'}$.

$$a_x = \frac{du_x}{dt} = \frac{du_x}{dt'}\frac{dt'}{dt} = \frac{\left(1 + \dfrac{u_{x'}v}{c^2}\right) a_{x'} - (u_{x'} + v)\dfrac{va_{x'}}{c^2}}{\left(1 + \dfrac{u_{x'}v}{c^2}\right)^2} \left(\frac{dt'}{dt}\right)$$

$$= \frac{a_{x'} + \dfrac{u_{x'}va_{x'}}{c^2} - \dfrac{u_{x'}va_{x'}}{c^2} - \dfrac{v^2 a_{x'}}{c^2}}{\left(1 + \dfrac{u_{x'}v}{c^2}\right)^2} \left(\frac{dt'}{dt}\right)$$

$$= \frac{a_{x'}\left(1 - \dfrac{v^2}{c^2}\right)}{\left(1 + \dfrac{vu_{x'}}{c^2}\right)^2} \left(\frac{dt'}{dt}\right)$$

Now substitute $\dfrac{1}{\gamma^2}$ for $1 - \dfrac{v^2}{c^2}$.

(2)

$$\frac{du_x}{dt} = \frac{a_{x'}}{\gamma^2 \left(1 + \dfrac{u_{x'}v}{c^2}\right)^2} \left(\frac{dt'}{dt}\right)$$

We will return to this equation later, after we work out the value of the time derivative on the right hand side.

For the y-component of the acceleration, begin with Equation 25.47 on page 1172 of the text:

$$u_y = \frac{u_{y'}}{\gamma\left(1 + \frac{vu_{x'}}{c^2}\right)}$$

Differentiate with respect to t, again using the chain and quotient rules.

$$a_y = \frac{du_y}{dt} = \frac{du_y}{dt'}\frac{dt'}{dt} = \frac{1}{\gamma}\left(\frac{\left(1 + \frac{vu_{x'}}{c^2}\right)a_{y'} - u_{y'}\frac{va_{x'}}{c^2}}{\left(1 + \frac{vu_{x'}}{c^2}\right)^2}\right)\left(\frac{dt'}{dt}\right).$$

So,

$$(3) \qquad \frac{du_y}{dt} = \frac{\left(1 + \frac{vu_{x'}}{c^2}\right)a_{y'} - u_{y'}\frac{va_{x'}}{c^2}}{\gamma\left(1 + \frac{vu_{x'}}{c^2}\right)^2}\left(\frac{dt'}{dt}\right)$$

In order to complete the transformation equations between $\vec{a}$ and $\vec{a'}$, we need an expression for $\frac{dt'}{dt}$. We start by differentiating the Lorentz transformation equation, Equation 25.36 on page 1164 of the text, which expresses t' as a function of t.

$$t' = \gamma\left(t - \frac{v}{c^2}x\right) \implies \frac{dt'}{dt} = \gamma\left(1 - \frac{v}{c^2}u_x\right).$$

Our goal, however, is to express a_x and a_y entirely in terms of the primed variables measured in S'. Therefore, we use equation (1) above to replace u_x in our last equation for $\frac{dt'}{dt}$.

$$\frac{dt'}{dt} = \gamma\left(1 - \frac{v}{c^2}\left(\frac{u_{x'} + v}{1 + \frac{u_{x'}v}{c^2}}\right)\right)$$

$$= \gamma\left(\frac{1 + \frac{u_{x'}v}{c^2} - \frac{u_{x'}v}{c^2} - \frac{v^2}{c^2}}{1 + \frac{u_{x'}v}{c^2}}\right)$$

$$= \gamma\left(\frac{1 - \frac{v^2}{c^2}}{1 + \frac{u_{x'}v}{c^2}}\right)$$

$$= \gamma\left(\frac{\frac{1}{\gamma^2}}{1 + \frac{u_{x'}v}{c^2}}\right) = \frac{1}{\gamma}\left(\frac{1}{1 + \frac{u_{x'}v}{c^2}}\right)$$

Substitute this expression for $\frac{dt'}{dt}$ into equation (2) above, to find

$$a_x = \frac{d}{dt}u_x = \frac{a_{x'}}{\gamma^2\left(1 + \frac{u_{x'}v}{c^2}\right)^2}\frac{1}{\gamma}\left(\frac{1}{1 + \frac{u_{x'}v}{c^2}}\right) = \frac{a_{x'}}{\gamma^3\left(1 + \frac{u_{x'}v}{c^2}\right)^3},$$

which is what we wanted to show for a_x.

Similarly, for a_y, substitute the last expression for $\dfrac{dt'}{dt}$ into equation (3) above, to find

$$a_y = \frac{du_y}{dt} = \frac{\left(1 + \frac{vu_{x'}}{c^2}\right)a_{y'} - u_{y'}\frac{va_{x'}}{c^2}}{\gamma\left(1 + \frac{vu_{x'}}{c^2}\right)^2}\left(\frac{1}{\gamma}\left(\frac{1}{1 + \frac{u_{x'}v}{c^2}}\right)\right) = \frac{\left(1 + \frac{u_{x'}v}{c^2}\right)a_{y'} - \left(\frac{u_{y'}v}{c^2}\right)a_{x'}}{\gamma^2\left(1 + \frac{u_{x'}v}{c^2}\right)^3}.$$

Curiously, acceleration and its transformation equations rarely are used or needed in special relativity. In other words, although acceleration plays a central role in classical mechanics (because of the central role of Newton's Second Law $\vec{\mathbf{F}} = m\vec{\mathbf{a}}$), it plays almost no role in special relativity because problems involving relativistic dynamics almost always can be solved using conservation of momentum and energy. Consequently, very few books even *mention* the transformation equations for acceleration. Two that discuss it in some detail are: A.P. French, *Special Relativity*, (W.W. Norton, NY, 1968) and H. Arzelies, *Relativistic Point Dynamics*, (Pergamon Press, Oxford, 1972).

25.37   Since the firefly is at rest in S and at the origin, events corresponding to successive flashes at intervals of τ_0 are:

Event 0. $x_0 = 0$ m   and $t_0 = 0$ s.

Event 1. $x_1 = 0$ m   and $t_1 = \tau_0$.

Event 2. $x_2 = 0$ m   and $t_2 = 2\tau_0$.

Event 3. $x_3 = 0$ m   and $t_3 = 3\tau_0$.

Use the Lorentz transformation equations to transform the coordinates of these events to the frame S' of the detector, which is moving away from the firefly at speed v. The transformation equations (from Equations 25.33 and 25.36 on page 1164 of the text) are

$$x' = \gamma(x - vt) \qquad \text{and} \qquad t' = \gamma\left(t - \frac{v}{c^2}x\right).$$

Here the events are in the S' frame:

Event 0. $x_0' = 0$ m   and $t_0' = 0$ s

Event 1. $x_1' = -\gamma v\tau_0$ and $t_1' = \gamma\tau_0$

Event 2. $x_2' = -2\gamma v\tau_0$ and $t_2' = 2\gamma\tau_0$

Event 3. $x_3' = -3\gamma v\tau_0$ and $t_3' = 3\gamma\tau_0$

Event 1 occurs when the clocks in S' read $\gamma\tau_0$, but at a distance $\gamma v\tau_0$ from the detector. The light must propagate over this distance at the speed c to the detector. Hence the length of time τ between detection of light from Event 0 and Event 1 is

$$\tau = \gamma\tau_0 + \frac{\gamma v\tau_0}{c} = \tau_0\gamma\left(1 + \frac{v}{c}\right).$$

25.41.

The frequency ν is the reciprocal of the period, so

$$\nu = \frac{1}{\tau} = \frac{1}{\tau_0 \gamma \left(1 + \dfrac{v}{c}\right)}$$

$$= \frac{1}{\left(\dfrac{1}{\nu_0}\right)\left(\dfrac{1}{\sqrt{1 - \dfrac{v^2}{c^2}}}\right)\left(1 + \dfrac{v}{c}\right)}$$

$$= \nu_0 \frac{\sqrt{1 - \dfrac{v^2}{c^2}}}{1 + \dfrac{v}{c}}$$

$$= \nu_0 \frac{\sqrt{\left(1 - \dfrac{v}{c}\right)\left(1 + \dfrac{v}{c}\right)}}{1 + \dfrac{v}{c}}$$

$$= \nu_0 \sqrt{\frac{1 - \dfrac{v}{c}}{1 + \dfrac{v}{c}}}$$

Notice, from this formula, that $\nu < \nu_0$, so the change in frequency is a redshift.

25.41 The frequencies in the transverse Doppler effect are related by Equation 25.60 on page 1179 of the text,

$$\nu = \nu_0 \sqrt{1 - \frac{v^2}{c^2}}.$$

The problem is given to us in terms of wavelengths rather than frequencies, so we'll begin by rewriting this relation in terms of wavelengths λ and λ_0. Since $c = \nu\lambda$, we have $\nu = \dfrac{c}{\lambda}$, and also $\nu_0 = \dfrac{c}{\lambda_0}$. Hence,

$\dfrac{c}{\lambda} = \dfrac{c}{\lambda_0}\sqrt{1 - \dfrac{v^2}{c^2}}$, so

(1)
$$\frac{\lambda_0}{\lambda} = \sqrt{1 - \frac{v^2}{c^2}}.$$

We'll first solve this equation for v, and then substitute the known values of $\lambda_0 = 656.282$ nm, $\lambda = 656.382$ nm, and $c = 3.00 \times 10^8$ m/s.

Begin by squaring both sides of (1), and then turn the crank.

$$\left(\frac{\lambda_0}{\lambda}\right)^2 = 1 - \frac{v^2}{c^2} \implies \frac{v^2}{c^2} = 1 - \left(\frac{\lambda_0}{\lambda}\right)^2 \implies$$

$$v = \left(\sqrt{1 - \left(\frac{\lambda_0}{\lambda}\right)^2}\right)c = \left(\sqrt{1 - \left(\frac{656.282 \text{ nm}}{656.382 \text{ nm}}\right)^2}\right)c = (0.0175)c = 5.25 \times 10^6 \text{ m/s}.$$

25.45 Say that your mass is 60 kg, then your rest energy is

$$E_{\text{rest}} = mc^2 = (60 \text{ kg})(3.00 \times 10^8)^2 = 5.4 \times 10^{18} \text{ J}$$

$$= (5.4 \times 10^{18} \text{ W·s})\left(\frac{\text{kW}}{10^3 \text{ W}}\right)\left(\frac{\text{h}}{3600 \text{ s}}\right) = 1.5 \times 10^{12} \text{ kW·h}.$$

At a cost of \$0.10 per kilowatt-hour, the price would be

$$(1.5 \times 10^{12} \text{ kW·h}) \left(\frac{\$0.10}{\text{kW·h}} \right) = \$1.5 \times 10^{11}.$$

That's 150 *billion* dollars!

25.49 First find the relativistic factor

$$\gamma = \frac{1}{\sqrt{1 - \dfrac{v^2}{c^2}}} = \frac{1}{\sqrt{1 - (0.600)^2}} = 1.25$$

The kinetic energy is

$$\text{KE} = (\gamma - 1)mc^2 = (1.25 - 1)(9.11 \times 10^{-31} \text{ kg})(3.00 \times 10^8 \text{ m/s})^2 = 2.05 \times 10^{-14} \text{ J}.$$

The total relativistic energy is

$$E = \gamma mc^2 = 1.25(9.11 \times 10^{-31} \text{ kg})(3.00 \times 10^8 \text{ m/s})^2 = 1.02 \times 10^{-13} \text{ J}.$$

The magnitude of the momentum is

$$p = \gamma mv = 1.25(9.11 \times 10^{-31} \text{ kg})0.600(3.00 \times 10^8 \text{ m/s}) = 2.05 \times 10^{-22} \text{ kg·m/s}.$$

As a check, evaluate both sides of the equation

$$E^2 = p^2 c^2 + m^2 c^4.$$

The left hand side is

$$\text{LHS} = (1.02 \times 10^{-13} \text{ J})^2 = 1.04 \times 10^{-26} \text{ J}^2.$$

The right hand side is

$$\text{RHS} = (2.05 \times 10^{-22} \text{ kg·m/s})^2 (3.00 \times 10^8 \text{ m/s})^2 + (9.11 \times 10^{-31} \text{ kg})^2 (3.00 \times 10^8 \text{ m/s})^4 = 1.05 \times 10^{-26} \text{ J}^2.$$

Close enough!

25.53

a) The condition is

$$E = 1.10 E_{\text{rest}} \implies \gamma mc^2 = 1.10 mc^2 \implies \gamma = 1.10 \implies \frac{1}{\sqrt{1 - \dfrac{v^2}{c^2}}} = 1.10.$$

Square both sides of the last equation and solve for v.

$$\frac{1}{1 - \dfrac{v^2}{c^2}} = (1.10)^2 \implies 1 - \frac{v^2}{c^2} = \frac{1}{(1.10)^2} \implies v = \left(\sqrt{1 - \frac{1}{(1.10)^2}} \right) c = 0.417c = 1.25 \times 10^8 \text{ m/s}.$$

b) By the CWE theorem, the work done is equal to the change in the kinetic energy:

$$W = \Delta \text{KE} = \text{KE}_{\text{f}} - \text{KE}_{\text{i}} = (\gamma - 1)mc^2 - 0 \text{ J}$$

$$= (1.10 - 1)(1.67 \times 10^{-27} \text{ kg})(3.00 \times 10^8 \text{ m/s})^2 = 1.5 \times 10^{-11} \text{ J}.$$

Note that when expressed in units of electron-volts,

$$W = (1.5 \times 10^{-11} \text{ J}) \left(\frac{\text{eV}}{1.602 \times 10^{-19} \text{ J}} \right) = 9.4 \times 10^7 \text{ eV}.$$

So, to reach this kinetic energy, the proton would have to be accelerated through a potential difference of 94 million volts.

25.57

a) First we'll find γ, and then we'll use it to find v.
The kinetic energy is

$$\text{KE} = (\gamma - 1)mc^2 \implies$$

$$\gamma = 1 + \frac{\text{KE}}{mc^2} = 1 + \frac{1.60 \times 10^{-15} \text{ J}}{(9.11 \times 10^{-31} \text{ kg})(3.00 \times 10^8 \text{ m/s})^2} = 1.0195$$

Now find v,

$$\gamma = 1.0195 \implies \frac{1}{\sqrt{1 - \dfrac{v^2}{c^2}}} = 1.0195 \implies \sqrt{1 - \frac{v^2}{c^2}} = \frac{1}{1.0195} \implies 1 - \frac{v^2}{c^2} = \left(\frac{1}{1.0195}\right)^2$$

$$\implies v = \sqrt{1 - \left(\frac{1}{1.0195}\right)^2} \, c = 0.1947c = 5.841 \times 10^7 \text{ m/s}.$$

b) We're given that $\text{KE} = 1.60 \times 10^{-15} \text{ J}$, so

$$\text{KE} = (1.60 \times 10^{-15} \text{ J}) \left(\frac{\text{eV}}{1.602 \times 10^{-19} \text{ J}}\right) = 9.99 \times 10^3 \text{ eV} = 9.99 \text{ keV}.$$

c) The magnitude of the momentum is

$$p = \gamma m v.$$

We computed $\gamma = 1.0195$ and $v = 5.841 \times 10^7 \text{ m/s}$ in part a), so

$$p = 1.0195(9.11 \times 10^{-31} \text{ kg})(5.841 \times 10^7 \text{ m/s}) = 5.42 \times 10^{-23} \text{ kg·m/s}.$$

25.61

a) The kinetic energy is defined as

$$\text{KE} = (\gamma - 1)mc^2.$$

So we'll start (as usual in these problems!) by computing the relativistic factor γ.

$$\gamma = \frac{1}{\sqrt{1 - \dfrac{v^2}{c^2}}} = \frac{1}{\sqrt{1 - (0.980)^2}} = 5.0.$$

Then

$$\text{KE} = (\gamma - 1)mc^2 = (5.0 - 1)(70.0 \text{ kg})(3.00 \times 10^8 \text{ m/s})^2 = 2.5 \times 10^{19} \text{ J}.$$

b) The politician is at rest, in his own frame, so according to him his kinetic energy is 0 J.

c) For clocks on Earth the trip time is the distance divided by the speed. When light years are the unit of distance, it is most convenient to use c as the unit of speed, and years as the unit of time. In these units, $c = 1 \text{ LY/y}$. Thus the time required is

$$\frac{\text{distance}}{\text{speed}} = \frac{20.0 \text{ LY}}{0.980c} = \frac{20.0}{0.980} \text{ y} = 20.4 \text{ y}.$$

d) According to the politician, the distance between the Earth and the planet is contracted to

$$\ell = \frac{\ell_0}{\gamma} = \frac{20.0 \text{ LY}}{5.0} = 4.0 \text{ LY}.$$

Hence, on arrival at the planet, the Earth is only 4.0 LY away — according to the politician.

e)   The politician measures the length of time for the trip to be

$$\frac{\text{distance}}{\text{speed}} = \frac{4.0 \text{ LY}}{0.980c} = 4.1 \text{ y}.$$

In summary, the politician concludes that he traveled 4.0 LY in 4.1 y, but Earth bound creatures believe the journey was 20.0 LY, and that it took the politician 20.4 y.

Suppose that the politician quickly turns around and comes home at the same speed — perhaps by using his head to bounce off the planet with a completely elastic collision, and then firing off a rocket or two to make up for any momentum imparted to the planet. Then when the politician arrives back on Earth, he will have aged only 8.2 y, while his Earth bound opposition will be 40.8 y older.

There are a number of subtleties involved here, which often are discussed as "the twin paradox." See, for example:

Gerald Holton, "AAPT Resource Letter SRT on Special Relativity Theory, *American Journal of Physics, 30, #6*, pages 4

Richard A. Muller, "The Twin Paradox in Special Relativity," *American Journal of Physics, 40, #7*, pages 966-969 (July, 1972);

Margaret Stautberg Greenwood, "Use of Doppler-shifted Light Beams to Measure Time During Acceleration," *American Journal of Physics, 44, #3*, pages 259-263 (March, 1976);

Donald E. Hall, "Intuition, Time Dilation and the Twin Paradox," *The Physics Teacher, 16, #4*, pages 209-215 (April, 1978); and

Robert H. Good, "Uniformly Accelerated Reference Frames and the Twin Paradox," *American Journal of Physics, 50, #3*, pages 232-238 (March, 1982).

Also, note that there are many *experiments* confirming time dilation and the twin "paradox" (which, of course, isn't *really* a paradox). See, for example, the discussions in:

Vernon D. Barger and Martin G. Olsson, *Classical Mechanics: A Modern Perspective*, pages 350-353, (McGraw-Hill, New York, 1995);

John R. Taylor and Chris D. Zafiratos, *Modern Physics for Scientists and Engineers*, Chapter Two (Prentice-Hall, New York, 1991); and

Ralph Baierlein, *Newton to Einstein: The Trail of Light* (Cambridge University Press, New York, 1992).

25.65

a)   We want to show that

(1) $$\frac{\gamma^2 v^2}{\gamma + 1} = c^2(\gamma - 1).$$

Notice that the right-hand side of (1) depends just on γ and c, and doesn't involve v at all. This suggests that we first express v^2 in terms of γ and c, then substitute this expression for v^2 into the left-hand side of (1), and hope that after simplification we end up with the right-hand side of (1). That's the plan. So first try to write v in terms of γ and c.

Start with the definition of γ and solve for v^2.

$$\gamma = \frac{1}{\sqrt{1 - \frac{v^2}{c^2}}} \implies \gamma^2 = \frac{1}{1 - \frac{v^2}{c^2}} \implies 1 - \frac{v^2}{c^2} = \frac{1}{\gamma^2} \implies v^2 = c^2(1 - \frac{1}{\gamma^2}) = c^2\left(\frac{\gamma^2 - 1}{\gamma^2}\right).$$

Good. That gives us an expression for v^2 in terms of γ and c.

Now substitute the expression for v^2 into the left-hand side of (1) and hope for the best:

$$\begin{aligned}
\frac{\gamma^2 v^2}{\gamma + 1} &= \frac{\gamma^2 c^2 \left(\frac{\gamma^2 - 1}{\gamma^2} \right)}{\gamma + 1} \\
&= \frac{c^2 \left(\gamma^2 - 1 \right)}{\gamma + 1} \\
&= \frac{c^2 (\gamma - 1)(\gamma + 1)}{\gamma + 1} \\
&= c^2 (\gamma - 1).
\end{aligned}$$

Hooray for our side!

b)

$$E = \gamma m c^2 = mc^2 + mc^2 (\gamma - 1) = mc^2 + m \frac{\gamma^2 v^2}{\gamma + 1}.$$

c)   From part b), $E = mc^2 + m \dfrac{\gamma^2 v^2}{\gamma + 1}$. But we also know that $E = mc^2 + \text{KE}$, hence

(1)
$$\text{KE} = m \frac{\gamma^2 v^2}{\gamma + 1}.$$

Now $\text{KE}_{\text{class}} = \frac{1}{2} mv^2$, so $mv^2 = 2\text{KE}_{\text{class}}$. Substitute this expression for mv^2 into (1) and simplify:

(2)
$$\text{KE} = \text{KE}_{\text{class}} \frac{2\gamma^2}{\gamma + 1},$$

which is what we wanted to show.

Notice that we can rewrite (2) as

(3)
$$\text{KE} = 2\gamma \left(\frac{\gamma}{\gamma + 1} \right) \text{KE}_{\text{class}}.$$

Since $1 \le \gamma < \infty$, we see from (3) that

(4)
$$\gamma \text{KE}_{\text{class}} \le \text{KE} < 2\gamma \text{KE}_{\text{class}}.$$

When $v \ll c$, then γ is close to 1 and therefore $2\gamma \left(\dfrac{\gamma}{\gamma + 1} \right)$ is close to γ. When v is close to c, then γ is very large so $2\gamma \left(\dfrac{\gamma}{\gamma + 1} \right)$ is close to 2γ. Therefore (3) tells us that

(5)
$$\text{KE} \approx \begin{cases} \gamma \text{KE}_{\text{class}} & \text{for } v \ll c, \text{ and} \\ 2\gamma \text{KE}_{\text{class}} & \text{for } v \text{ close to } c. \end{cases}$$

These approximations are useful if you want a quick guesstimate, if you want to do a quick check on a more detailed computation, or if you don't have your calculator with you.

25.69   Here's the picture.

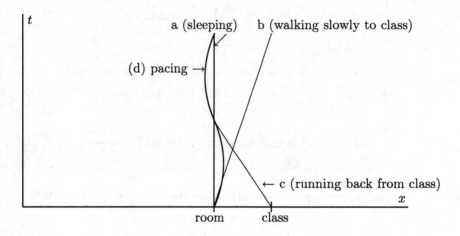

We've started each world line at the same time — $t = 0$ s. We've started all but one at the same point in space — $x =$ your room. We've started the one for returning from class at the class.

a)   While sleeping, you do not change your position, so the graph is a vertical straight line.

b)   The inverse of the slope on the space-time diagram is your velocity component, so walking to class is a line with a steep slope. We assume class is to the right of your room.

c)   Running from class (back to your room) is a line with slope less in magnitude than that in b), but of the opposite sign.

d)   Pacing back and forth is an oscillatory graph as shown above.

25.73   For every second along the vertical t axis, the light travels 3.00×10^8 m. The graph is a straight line with zero intercept and slope $\dfrac{1 \text{ s}}{3.00 \times 10^8 \text{ m}}$ Here's the graph.

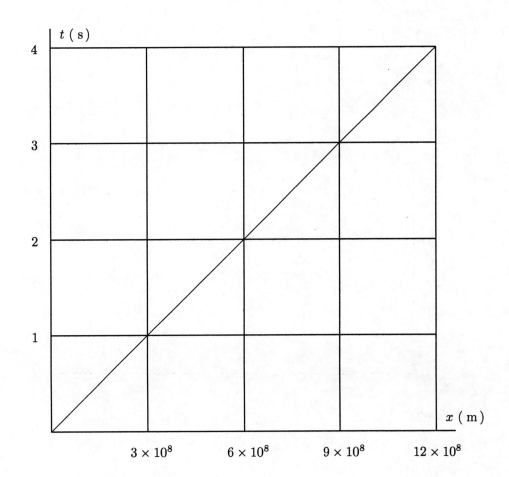

Chapter 26

An Aperitif: Modern Physics

26.1   The peak wavelength λ_{max} and temperature T are related by Wien's displacement law, Equation 26.1 on page 1210 of the text:

$$\lambda_{max}T = 0.28978 \times 10^{-2} \text{ m·K}.$$

Hence, if you double T, the peak wavelength is half its previous value.

26.5   Use Wien's displacement law, Equation 26.1 on page 1210 of the text:

$$\lambda_{max}T = 0.28978 \times 10^{-2} \text{ m·K} \implies T = \frac{0.28978 \times 10^{-2} \text{ m·K}}{\lambda_{max}} = \frac{0.28978 \times 10^{-2} \text{ m·K}}{6.5 \times 10^{-7} \text{ m}} = 4.5 \times 10^3 \text{ K}.$$

It is interesting to compare this with our Sun. The peak of the Sun's blackbody spectrum occurs at about $\lambda_{Sun} = 550 \text{ nm}$, so

$$T_{Sun} = \frac{0.28978 \times 10^{-2} \text{ m·K}}{5.5 \times 10^{-7} \text{ m}} = 5.3 \times 10^3 \text{ K}.$$

This is one of the reasons we believe Antares is a little cooler than our Sun.

26.9   The energy of a single photon is

$$E = h\nu.$$

Since $c = \nu\lambda$, we may rewrite this as

$$E = \frac{hc}{\lambda}.$$

For $\lambda = 400 \text{ nm}$, this gives us

$$E_{400 \text{ nm}} = \frac{(6.626 \times 10^{-34} \text{ J·s})(3.00 \times 10^8 \text{ m/s})}{400 \times 10^{-9} \text{ m}} = 4.97 \times 10^{-19} \text{ J}.$$

Expressed in electron-volts, this is

$$E_{400 \text{ nm}} = (4.97 \times 10^{-19} \text{ J}) \left(\frac{\text{eV}}{1.602 \times 10^{-19} \text{ J}} \right) = 3.10 \text{ eV}.$$

For $\lambda = 700 \text{ nm}$, we have

$$E_{700 \text{ nm}} = \frac{(6.626 \times 10^{-34} \text{ J·s})(3.00 \times 10^8 \text{ m/s})}{700 \times 10^{-9} \text{ m}} = 2.84 \times 10^{-19} \text{ J}.$$

Expressed in electron-volts, this is

$$E_{700 \text{ nm}} = (2.84 \times 10^{-19} \text{ J}) \left(\frac{\text{eV}}{1.602 \times 10^{-19} \text{ J}} \right) = 1.77 \text{ eV}.$$

Hence the energy range is 1.77 eV to 3.10 eV per photon.

26.13

a)   The energy of an incoming photon is

$$E = h\nu.$$

Since $c = \nu\lambda$, we may rewrite this as

$$E = \frac{hc}{\lambda}.$$

For $\lambda = 200$ nm, this gives us

$$E_{200\ \text{nm}} = \frac{(6.626 \times 10^{-34}\ \text{J·s})(3.00 \times 10^8\ \text{m/s})}{200 \times 10^{-9}\ \text{m}} = 9.94 \times 10^{-19}\ \text{J}.$$

Expressed in electron-volts, this is

$$E_{400\ \text{nm}} = (9.94 \times 10^{-19}\ \text{J})\left(\frac{\text{eV}}{1.602 \times 10^{-19}\ \text{J}}\right) = 6.20\ \text{eV}.$$

Now use the equation for the photoelectric effect:

$$h\nu = W + \text{KE}_{\text{max}} \implies \text{KE}_{\text{max}} = h\nu - W = 6.20\ \text{eV} - 4.1\ \text{eV} = 2.1\ \text{eV}.$$

b)   The stopping potential in volts is numerically equal to the maximum kinetic energy in electron-volts:

$$V_s = \frac{2.1\ \text{eV}}{e} = 2.1\ \text{V}.$$

c)   The cutoff wavelength λ_c may be found from the work function.

$$W = h\nu_c = \frac{hc}{\lambda_c} \implies \lambda_c = \frac{hc}{W}$$

$$\implies W = \frac{(6.626 \times 10^{-34}\ \text{J·s})(3.00 \times 10^8\ \text{m/s})}{4.1\ \text{eV}}\left(\frac{\text{eV}}{1.602 \times 10^{-19}\ \text{J}}\right) = 3.0 \times 10^{-7}\ \text{m} = 3.0 \times 10^2\ \text{nm}.$$

26.17

a)   Determine the cutoff frequency from the work function.

$$W = h\nu_c \implies \nu_c = \frac{W}{h} = \frac{2.0\ \text{eV}}{(6.626 \times 10^{-34}\ \text{J·s})}\left(\frac{1.602 \times 10^{-19}\ \text{J}}{\text{eV}}\right) = 4.8 \times 10^{14}\ \text{Hz}.$$

The incident frequency is greater than the cutoff frequency, so photoelectrons will be released. Thus, the photoelectric effect will occur.

b)   From the equation for the photoelectric effect, $h\nu = W + \text{KE}_{\text{max}}$, we have

$$\text{KE}_{\text{max}} = h\nu - W.$$

This is the equation we want to graph. The graph of KE_{max} versus ν is a straight line. It takes only two points to determine a straight line. One of the points on the line, is the intercept with the horizontal axis, where $\text{KE}_{\text{max}} = 0$ eV. This occurs at the cutoff frequency, $\nu_c = 4.8 \times 10^{14}$ Hz found in part a). To find another point on the line, arbitrarily let $\nu = 6.0 \times 10^{14}$ Hz. The corresponding KE_{max}, measured in eV, is

$$\text{KE}_{\text{max}} = h(6.0 \times 10^{14}\ \text{Hz}) - W$$

$$= (6.626 \times 10^{-34}\ \text{J·s})(6.0 \times 10^{14}\ \text{Hz})\left(\frac{\text{eV}}{1.602 \times 10^{-19}\ \text{J}}\right) - 2.0\ \text{eV} = 2.5\ \text{eV} - 2.0\ \text{eV} = 0.5\ \text{eV}.$$

Now just draw the straight line joining $(4.8 \times 10^{14}\ \text{Hz}, 0\ \text{eV})$ to $(6.0 \times 10^{14}\ \text{Hz}, 0.5\ \text{eV})$. Here's the result.

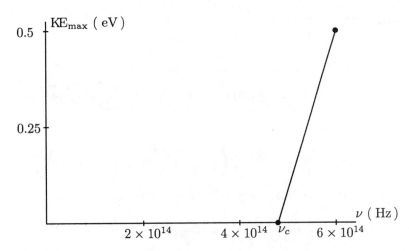

c)   Neither ν, nor KE_{max} depends upon the light intensity, so the graph is unchanged by changes in intensity.

26.21

a)   From Equation 26.24 on page 1220 of the text, the orbital radius is

$$r = \frac{n^2 \hbar^2 (4\pi\epsilon_0)}{mZe^2} = \frac{n^2 \hbar^2}{\dfrac{1}{4\pi\epsilon_0} mZe^2}$$

$$= \frac{2^2 (1.055 \times 10^{-34} \text{ J·s})^2}{(9.00 \times 10^9 \text{ N·m}^2/\text{C}^2)(9.11 \times 10^{-31} \text{ kg})(1)(1.602 \times 10^{-19} \text{ C})^2} = 2.12 \times 10^{-10} \text{ m}.$$

b)   The orbital speed is found from the magnitude of the orbital angular momentum.

$$L = n\hbar \implies mvr = n\hbar \implies v = \frac{n\hbar}{mr} = \frac{2(1.055 \times 10^{-34} \text{ J·s})}{(9.11 \times 10^{-31} \text{ kg})(2.12 \times 10^{-10} \text{ m})} = 1.09 \times 10^6 \text{ m/s}.$$

Notice that although this is quite speedy, it is still only about 0.36% of the speed of light, that is,

$$\frac{1.09 \times 10^6 \text{ m/s}}{3.00 \times 10^8 \text{ m/s}} = 0.00363.$$

c)   The orbital angular momentum is

$$L = n\hbar = 2(1.055 \times 10^{-34} \text{ J·s}) = 2.110 \times 10^{-34} \text{ kg·m}^2/\text{s}.$$

d)   The magnitude of the centripetal acceleration is

$$a_{centripetal} = \frac{v^2}{r} = \frac{(1.09 \times 10^6 \text{ m/s})^2}{2.12 \times 10^{-10} \text{ m}} = 5.60 \times 10^{21} \text{ m/s}^2.$$

e)   The ratio is

$$\frac{a_{centripetal}}{g} = \frac{5.60 \times 10^{21} \text{ m/s}^2}{9.81 \text{ m/s}^2} = 5.71 \times 10^{20}.$$

WOW!

f)   Because $\dfrac{v}{c}$ is only 0.00363, we may use the classical (nonrelativistic) expression for the kinetic energy:

$$KE = \frac{1}{2}mv^2 = \frac{1}{2}(9.11 \times 10^{-31} \text{ kg})(1.09 \times 10^6 \text{ m/s})^2 = 5.41 \times 10^{-19} \text{ J}.$$

Measured in electron-volts, this is

$$KE = (5.41 \times 10^{-19} \text{ J}) \left(\frac{eV}{1.602 \times 10^{-19} \text{ J}} \right) = 3.38 \text{ eV}.$$

g)   The potential energy of the electron is

$$PE = qV = (-e)\left(\frac{1}{4\pi\epsilon_0}\right)\frac{e}{r} = -\left(\frac{1}{4\pi\epsilon_0}\right)\frac{e^2}{r}$$

$$= -(9.00 \times 10^9 \text{ N·m}^2/\text{C}^2)\frac{(1.602 \times 10^{-19} \text{ C})^2}{2.12 \times 10^{-10} \text{ m}} = -1.09 \times 10^{-18} \text{ J}$$

Measured in electron-volts, this is

$$PE = (-1.09 \times 10^{-18} \text{ J})\left(\frac{\text{eV}}{1.602 \times 10^{-19} \text{ J}}\right) = -6.80 \text{ eV}.$$

As a check, note that from Equation 26.25 on page 1220 of the text,

$$\frac{1}{2}mv^2 = \frac{1}{2}\left(\frac{1}{4\pi\epsilon_0}\frac{Ze^2}{r}\right) = \frac{1}{2}(-PE) \implies KE = \frac{1}{2}(-PE).$$

Therefore, since $E = KE + PE$, we have

$$PE = E - KE = \frac{-13.6 \text{ eV}}{2^2} - \frac{1}{2}(-PE) = \frac{-13.6 \text{ eV}}{4} + \frac{PE}{2} \implies$$

$$\frac{PE}{2} = \frac{-13.6 \text{ eV}}{4} \implies PE = \frac{-13.6 \text{ eV}}{2} = -6.80 \text{ eV},$$

as expected.

26.25

a)   First find the energy $E_{\text{photon absorbed}}$ necessary to barely ionize the hydrogen atom. Next find the wavelength that will provide that energy.

The initial energy E_i of the atom, final energy E_f of the atom, and the energy E_{photon} of the absorbed photon satisfy Equation 26.32 (for absorption) on page 1221 of the text,

$$E_f = E_i + E_{\text{photon absorbed}} \implies E_{\text{photon absorbed}} = E_f - E_i.$$

From Equation 26.30 on page 1220, the energy levels of the hydrogen atom are

$$E_n = -\frac{13.6 \text{ eV}}{n^2}.$$

So, for the initial, $n = 2$ state,

$$E_i = E_2 = -\frac{13.6 \text{ eV}}{4} = -3.40 \text{ eV}.$$

In the final ionized state, $n = \infty$, so

$$E_f = E_\infty = 0 \text{ eV}.$$

Therefore,

$$E_{\text{photon absorbed}} = 0 \text{ eV} - (-3.40 \text{ eV}) = 3.40 \text{ eV}.$$

The energy of the absorbed photon is $h\nu = \frac{hc}{\lambda}$. Hence

$$\frac{hc}{\lambda} = 3.40 \text{ eV} \implies \lambda = \frac{hc}{3.40 \text{ eV}} = \frac{(6.626 \times 10^{-34} \text{ J·s})(3.00 \times 10^8 \text{ m/s})}{3.40 \text{ eV}}$$

$$= \frac{(6.626 \times 10^{-34} \text{ J·s})(3.00 \times 10^8 \text{ m/s})}{3.40 \text{ eV}}\left(\frac{\text{eV}}{1.602 \times 10^{-19} \text{ J}}\right) = 3.65 \times 10^{-7} \text{ m} = 365 \text{ nm}.$$

b)   The energy of a photon is $\dfrac{hc}{\lambda}$, so the longer the wavelength, the lower the energy. Hence, longer wavelength photons won't ionize a hydrogen atom initially in the $n = 2$ state.

26.29   The energy, measured in electron-volts, of a photon with wavelength $\lambda = 10$ nm is

$$E_{10\text{ nm}} = \frac{hc}{\lambda} = \frac{(6.626 \times 10^{-34}\text{ J·s})(3.00 \times 10^8\text{ m/s})}{10 \times 10^{-9}\text{ m}}$$

$$= \frac{(6.626 \times 10^{-34}\text{ J·s})(3.00 \times 10^8\text{ m/s})}{10 \times 10^{-9}\text{ m}} \left(\frac{\text{eV}}{1.602 \times 10^{-19}\text{ J}}\right) = 124\text{ eV}.$$

So, photons with wavelengths less than 10 nm have energies greater than 124 eV.

On the other hand, the energy of a photon emitted by going from state $n = n_i$ to state n_f is, from Equation 26.31 on page 1221 of the text,

$$E_{\text{emitted photon}} = E_i - E_f = E_{n_i} - E_{n_f} = \frac{-13.6\text{ eV}}{n_i^2} - \frac{-13.6\text{ eV}}{n_f^2} = (13.6\text{ eV})\left(\frac{1}{n_f^2} - \frac{1}{n_i^2}\right)$$

This quantity is at its biggest when $n_f = 1$ and $n_i = \infty$, and even then it is only 13.6 eV, which is less than 124 eV. Hence, there is no transition within the hydrogen atom that will produce x-rays.

26.33

a)   From Equation 26.48 on page 1227 of the text, the disintegration constant λ and half life $\tau_{1/2}$ are related by

$$\tau_{1/2} = \frac{\ln 2}{\lambda} \implies \lambda = \frac{\ln 2}{\tau_{1/2}} = \frac{\ln 2}{3.82\text{ d}} = 0.181\text{ d}^{-1}$$

b)   The amount N left after time t starting with an initial amount N_0 is

$$N = N_0 e^{-\lambda t} \implies \frac{N}{N_0} = e^{-\lambda t} = e^{-(0.181\text{ d}^{-1})(30\text{ d})} = 4.5 \times 10^{-3} = 0.45\%.$$

Actually radon itself is a relatively harmless gas. Scientists believe the harm to humans actually comes from one of the by-products of radon decay, polonium (Po). Radon undergoes α-decay

$$^{222}_{86}\text{Rn} \rightarrow {}^4_2\alpha + {}^{218}_{84}\text{Po}$$

with a half life of 3.82 days, but polonium undergoes α-decay with a half life of 3.10 *minutes*. Although radon is simply breathed in and out, polonium binds itself to lung tissue and then, within minutes, emits an α particle which can cause cell damage and cancer. So the danger comes not only from the α particles, but also from the relatively short half lives of the two α decays which release these α particles inside the lungs, and from the fact that polonium will not be breathed out like radon.

26.37   Since the N_0 inside the integral in Equation 26.55 on page 1228 clearly cancels with the N_0 in the denominator, the integration to be verified is

$$\int_{0\text{ s}}^{\infty\text{ s}} t\lambda e^{-\lambda t}\, dt = \frac{1}{\lambda}.$$

Because the upper limit of integration is ∞ s, this is an *improper* integral. Its value is determined by evaluating the integral with some finite time, b say, as the upper limit, and then taking the limit as $b \rightarrow \infty$ s. Therefore we need to show that

$$\lim_{b \rightarrow \infty\text{ s}} \int_{0\text{ s}}^{b} t\lambda e^{-\lambda t}\, dt = \frac{1}{\lambda}.$$

In order to show this, we first need an expression for $\int_{0\,s}^{b} t\lambda e^{-\lambda t}\, dt$.

In order to find *that*, let's first look at the indefinite integral $\int te^{-\lambda t}$. After we've found the indefinite integral, we'll tack on integration limits $0\,s$ and b and, finally, see what happens when $b \to \infty\,s$.

To evaluate the indefinite integral, we'll use the integration by parts formula: $\int u\, dv = uv - \int v\, du$, with $u = t$, $dv = \lambda e^{-\lambda t}\, dt$, and therefore $du = dt$, and $v = -e^{-\lambda t}$. Thus,

$$\int u\, dv = uv - \int v\, du$$

$$\implies \int te^{-\lambda t} = -te^{-\lambda t} + \int e^{-\lambda t}\, dt$$

$$= -te^{-\lambda t} - \frac{1}{\lambda}e^{-\lambda t}$$

Therefore, for any finite time b, we have

$$\int_{0\,s}^{b} t\lambda e^{-\lambda t}\, dt = \left(-te^{-\lambda t} - \frac{1}{\lambda}e^{-\lambda t} \right) \bigg|_{0\,s}^{b}$$

$$= \left(-be^{-\lambda b} - \frac{1}{\lambda}e^{-\lambda b} \right) - \left(-(0\,s)e^{-\lambda(0\,s)} - \frac{1}{\lambda}e^{-\lambda(0\,s)} \right)$$

$$= \left(-be^{-\lambda b} - \frac{1}{\lambda}e^{-\lambda b} \right) - \left(0\,s - \frac{1}{\lambda} \right)$$

$$= -be^{-\lambda b} - \frac{1}{\lambda}e^{-\lambda b} + \frac{1}{\lambda}$$

Now to see what happens to this last expression as $b \to \infty\,s$. The last term $\frac{1}{\lambda}$ stays put at $\frac{1}{\lambda}$. That's good. Now we just want the other terms to go to zero.

Look next at the second term

$$\frac{1}{\lambda}e^{-\lambda b} = \frac{1}{\lambda e^{\lambda b}}.$$

As $b \to \infty\,s$, the denominator $\lambda e^{\lambda b}$ goes to $\infty\,s^{-1}$, so the fraction $\frac{1}{\lambda e^{\lambda b}}$ goes to $0\,s$. That's good.

It remains to show that the first term,

$$be^{-\lambda b} = \frac{b}{e^{\lambda b}}$$

also goes to zero. This is actually an indeterminate form. As $b \to \infty\,s$, both the numerator and the denominator of $\frac{b}{e^{\lambda b}}$ go to infinity, so it's not immediately clear what happens to the fraction. Because the denominator is an exponential function of b, whereas the numerator is just a linear function, it seems likely that the denominator goes to infinity so much faster than the numerator, that the entire fraction should go to zero.

This intuition is borne out by l'Hôpital's rule. Since both numerator and denominator of $\frac{b}{e^{\lambda b}}$ go to infinity as b does, l'Hôpital's rule tells us that the limit of the fraction, if it exists, is the limit of the ratio of the derivatives of numerator and denominator. The derivative of the numerator is $\frac{db}{db} = 1$, and the derivative of the denominator is $\frac{d}{db}e^{\lambda b} = \lambda e^{\lambda b}$. Hence

$$\lim_{b \to \infty\,s} \frac{b}{e^{\lambda b}} = \lim_{b \to \infty\,s} \frac{1}{\lambda e^{\lambda b}} = 0\,s.$$

Hooray!

Summarizing,

$$\int_{0 \text{ s}}^{\infty \text{ s}} t\lambda e^{-\lambda t}\, dt = \lim_{b\to\infty \text{ s}} \int_{0 \text{ s}}^{b} t\lambda e^{-\lambda t}\, dt = \lim_{b\to\infty \text{ s}}\left(-be^{-\lambda b} - \frac{1}{\lambda}e^{-\lambda b} + \frac{1}{\lambda}\right) = 0\text{ s} + 0\text{ s} + \frac{1}{\lambda} = \frac{1}{\lambda}.$$

26.41 The disintegration constant is

$$\lambda = \frac{\ln 2}{\tau_{1/2}} = \frac{\ln 2}{5.0 \times 10^9 \text{ y}} = 1.4 \times 10^{-10} \text{ y}^{-1}.$$

Use Equation 26.47 on page 1227 of the text to compute the fraction $\frac{N}{N_0}$ that has not yet decayed:

$$N = N_0 e^{-\lambda t} \implies \frac{N}{N_0} e^{-\lambda t} = e^{-(1.4\times 10^{-10}\text{ y}^{-1})(2.5\times 10^9 \text{ y})} = 0.70 = 70\%.$$

26.45 Use the result of problem 26.44 on page 1228 of the text:

$$\frac{1}{\tau_{\text{effective}}} = \frac{1}{\tau} + \frac{1}{\tau_b} \implies \frac{1}{6.2 \text{ d}} = \frac{1}{8.1 \text{ d}} + \frac{1}{\tau_b} \implies \tau_b = \frac{(6.2 \text{ d})(8.1 \text{ d})}{8.1 \text{ d} - 6.2 \text{ d}} = 26 \text{ d}.$$

26.49

a) The disintegration constant is

$$\lambda = \frac{\ln 2}{\tau_{1/2}} = \frac{\ln 2}{5.73 \times 10^3 \text{ y}} = 1.21 \times 10^{-4} \text{ y}^{-1}.$$

b) The activity obeys Equation 26.52 on page 1228 of the text,

$$\left.\frac{dN}{dt}\right|_{\text{at time } t} = \left(\left.\frac{dN}{dt}\right|_{\text{at time 0 s}}\right) e^{-\lambda t}$$

$$\implies 750\left(\frac{\text{disintegrations}}{\text{hour}}\right) = 960\left(\frac{\text{disintegrations}}{\text{hour}}\right) e^{-\lambda t}$$

$$\implies 0.781 = e^{-\lambda t}.$$

Take logarithms of both sides of the last equation

$$\ln 0.781 = -\lambda t \implies t = \frac{\ln 0.781}{-\lambda} = \frac{\ln 0.781}{-1.21 \times 10^{-4} \text{ y}^{-1}} = 2.04 \times 10^3 \text{ y}.$$

So the lecture notes are a little over 2000 y old.

26.53 One cannot say with certainty, since the process is statistical. It is possible that all 5 may have decayed, or none. The *most likely* number remaining is 2 or 3.

26.57 Use Equation 26.78 on page 1234 of the text,

$$\Delta\lambda = \frac{h}{mc}(1 - \cos\theta) = \frac{6.626 \times 10^{-34} \text{ J·s}}{(9.11 \times 10^{-31})(3.00 \times 10^8 \text{ m/s})}(1 - \cos 90°) = 2.42 \times 10^{-12} \text{ m} = 0.00242 \text{ nm}$$

The wavelength of the scattered photon is

$$\lambda' = \lambda + \Delta\lambda = 0.1000 \text{ nm} + 0.00242 \text{ nm} = 0.1024 \text{ nm}.$$

26.61   If λ is the wavelength, then $\frac{1}{\lambda}$ is the number of wavelengths per unit length — the *wave number*. So we want to show that $p = h\frac{1}{\lambda}$. The magnitude of the momentum of a photon is

$$p = \frac{E}{c} = \frac{h\nu}{c} = h\frac{\nu}{c} = h\frac{1}{\lambda} = h(\text{wave number}).$$

Which was to be shown.

26.65   From the Bohr angular momentum postulate,

$$L = n\hbar,$$

we have

$$mvr = n\hbar \implies pr = n\hbar = n\frac{h}{2\pi} \implies 2\pi r = n\frac{h}{p} = n\lambda \implies n = \frac{2\pi r}{\lambda}$$

So the quantum number n, is the number of de Broglie wave lengths λ that are in the length $2\pi r$ of one Bohr orbit.

26.69

a)   Because $\dfrac{v}{c} = \dfrac{2.0 \times 10^6 \text{ m/s}}{3.00 \times 10^8 \text{ m/s}} = 0.0067 \ll 1$, we may use nonrelativistic mechanics to compute the momentum of the electron. Therefore the magnitude of its momentum is

$$p_{\text{electron}} = mv = (9.11 \times 10^{-31} \text{ kg})(2.0 \times 10^6 \text{ m/s}) = 1.8 \times 10^{-24} \text{ kg·m/s}.$$

The magnitude of the momentum of a photon is

$$p_{\text{photon}} = \frac{E}{c} = \frac{h\nu}{c} = \frac{h}{\lambda_{\text{photon}}} \implies \lambda_{\text{photon}} = \frac{h}{p_{\text{photon}}}.$$

So, since $p_{\text{photon}} = p_{\text{electron}} = 1.8 \times 10^{-24} \text{ kg·m/s}$,

$$\lambda_{\text{photon}} = \frac{6.626 \times 10^{-34} \text{ J·s}}{1.8 \times 10^{-24} \text{ kg·m/s}} = 3.7 \times 10^{-10} \text{ m} = 0.37 \text{ nm}.$$

b)   The de Broglie wavelength λ of *any* particle is $\dfrac{h}{p}$ — the magnitude of its momentum p divided into Planck's constant h. So particles whose momentums have the same magnitude also have the same de Broglie wavelength. In particular, the de Broglie wavelength of the electron in this problem is the same as for the photon,

$$\lambda_{\text{electron}} = \lambda_{\text{photon}} = 0.37 \text{ nm}.$$

Chapter 27

An Introduction to Quantum Mechanics

Introductory Remark Most of the problems in this chapter are applications of the time-energy or position-momentum uncertainty relations. They are unlike the problems in previous chapters in that you are not asked to make rigorous calculations *accurate* to a given number of significant figures. Rather, the purpose of these problems is to give order of magnitude *estimates* in the quickest and easiest way possible. So don't be alarmed if you are have to make "reasonable" assumptions which are often but not always true, if the use of mathematics is sometimes a little sloppy, or if the physical picture upon which your calculations are based is oversimplistic. Remember, we are only trying to *estimate*, You can find a good introduction to the importance and use of estimation problems in Section 1.7 on page 16 of the text.

27.1 Planck's constant expressed in electron-volt·seconds is

$$h = 6.626 \times 10^{-34} \text{ J·s} = (6.626 \times 10^{-34} \text{ J·s}) \left(\frac{\text{eV}}{1.602 \times 10^{-19} \text{ J}} \right) = 4.136 \times 10^{-15} \text{ eV·s}.$$

Therefore, $\hbar$ in electron-volts·seconds is

$$\hbar = \frac{h}{2\pi} = \frac{4.136 \times 10^{-15} \text{ eV}}{2\pi} = 6.583 \times 10^{-16} \text{ eV·s}.$$

27.5 Use the time-energy uncertainty principle:

$$\Delta E \, \Delta t \geq h \implies \Delta t \geq \frac{h}{\Delta E} = \frac{6.626 \times 10^{-34} \text{ J·s}}{120 \times 10^6 \text{ eV}} \left(\frac{\text{eV}}{1.602 \times 10^{-19} \text{ J}} \right) = 3.45 \times 10^{-23} \text{ s}.$$

27.9

a) The frequency of the light is

$$\nu = \frac{c}{\lambda} = \frac{3.00 \times 10^8 \text{ m/s}}{200 \times 10^{-9} \text{ m}} = 1.50 \times 10^{15} \text{ Hz}.$$

b) Use the time-energy uncertainty principle:

$$\Delta E \, \Delta t \geq h \implies \Delta E \geq \frac{h}{\Delta t}$$

For a photon, $E = h\nu$, so $\Delta E = h\Delta\nu$, therefore,

$$h\Delta\nu \geq \frac{h}{\Delta t} \implies \Delta\nu \geq \frac{1}{\Delta t} = \frac{1}{1.0 \times 10^{-9} \text{ s}} = 1.0 \times 10^9 \text{ Hz}.$$

c)   For a photon,

$$E = pc \implies \Delta E = \Delta p\, c \implies h\Delta\nu = \Delta p\, c.$$

From the position-momentum uncertainty principle,

$$h \le \Delta x\, \Delta p_x = \Delta x \frac{\Delta E}{c} = \Delta x \frac{h\,\Delta\nu}{c}$$

$$\implies \Delta x \ge \frac{c}{\Delta\nu} = \frac{3.00 \times 10^8 \text{ m/s}}{1.0 \times 10^9 \text{ Hz}} = 0.30 \text{ m}.$$

27.13

a)   Apply the position-momentum uncertainty principle with $\Delta x = 1.0 \times 10^{-14}$ m.

$$\Delta x\, \Delta p_x \ge h \implies \Delta p_x \ge \frac{h}{\Delta x} = \frac{6.626 \times 10^{-34} \text{ J·s}}{1.0 \times 10^{-14} \text{ m}} = 7 \times 10^{-20} \text{ kg·m/s}.$$

b)   The magnitude of the relativistic momentum of the electron is $p = \gamma m v$. Substitute $\dfrac{1}{\sqrt{1 - \dfrac{v^2}{c^2}}}$ for γ, and we have

(1)
$$p = \frac{mv}{\sqrt{1 - \dfrac{v^2}{c^2}}}.$$

Solve this equation to find v in terms of m, p, and c.

$$p = \frac{mv}{\sqrt{1 - \dfrac{v^2}{c^2}}} \implies p^2 - \frac{p^2}{c^2}v^2 = m^2 v^2 \implies \left(m^2 + \frac{p^2}{c^2}\right)v^2 = p^2 \implies v = \frac{p}{\sqrt{m^2 + \dfrac{p^2}{c^2}}}$$

This is an increasing function of p. Although we could perhaps show this by differentiating v with respect p and showing that the derivative is positive, its most easily seen from the physics: The bigger the momentum is, for fixed mass m, the bigger the speed v must be. We assume that the magnitude of the momentum is greater than the uncertainty of the momentum, so $p \ge \Delta p$. From part a), we also have $\Delta p \ge 7 \times 10^{-20}$ kg·m/s. Therefore,

$$v = \frac{p}{\sqrt{m^2 + \dfrac{p^2}{c^2}}} \ge \frac{\Delta p}{\sqrt{m^2 + \dfrac{(\Delta p)^2}{c^2}}} \ge \frac{7 \times 10^{-20} \text{ m}}{\sqrt{m^2 + \dfrac{(7 \times 10^{-20} \text{ m})^2}{c^2}}}$$

$$= \frac{7 \times 10^{-20} \text{ m}}{\sqrt{(9.11 \times 10^{-31} \text{ kg})^2 + \dfrac{(7 \times 10^{-20} \text{ kg·m/s})^2}{(3.00 \times 10^8 \text{ m/s})^2}}}$$

When we evaluate this mess (using a calculator!), we end up with

$$v = 299\,997\,713.5 \text{ m/s}.$$

To three significant figures (more than we're entitled to!), this is 3.00×10^8 m/s $= c$. Thus, we conclude that v is just under the speed of light.

c) Use the expression for the relativistic total energy:

$$E^2 = p^2 c^2 + m^2 c^4,$$

and use the uncertainty in p as a lower bound for p: $p \geq \Delta p = 7 \times 10^{-20}$ kg·m/s. Then

$$E = \sqrt{p^2 c^2 + m^2 c^4} \geq \sqrt{(\Delta p)^2 c^2 + m^2 c^4}$$
$$= \sqrt{(7 \times 10^{-20} \text{ kg·m/s})^2 (3.00 \times 10^8 \text{ m/s})^2 + (9.11 \times 10^{-31} \text{ kg})^2 (3.00 \times 10^8 \text{ m/s})^4} = 2 \times 10^{-11} \text{ J}.$$

In mega-electron-volts, this is

$$E = (2 \times 10^{-11} \text{ J}) \left(\frac{\text{eV}}{1.602 \times 10^{-19} \text{ J}} \right) \left(\frac{\text{MeV}}{10^6 \text{ eV}} \right) = 1 \times 10^2 \text{ MeV}.$$

This is approximately 2×10^2 times the rest energy of the electron.

This problem shows one reason why we don't expect to find electrons in the nucleus of an atom, or as basic constituents of the nucleus. Confining them to such a small volume would give them plenty of energy to escape.